Implementing Project and Program Benefit Management

Best Practices in Portfolio, Program, and Project Management

Series Editor
Ginger Levin

RECENTLY PUBLISHED TITLES

Implementing Project and Program Benefit Management
Kenn Dolan

Culturally Tuning Change Management
Risto Gladden

The Four Pillars of Portfolio Management: Organizational Agility, Strategy, Risk, and Resources
Olivier Lazar

Systems Engineering for Projects: Achieving Positive Outcomes in a Complex World
Lory Mitchell Wingate

The Human Factor in Project Management
Denise Thompson

Project Business Management
Oliver F. Lehmann

PgMP® Exam Test Preparation: Test Questions, Practice Tests, and Simulated Exams
Ginger Levin

Managing Complex Construction Projects: A Systems Approach
John K. Briesemeister

Managing Project Competence: The Lemon and the Loop
Rolf Medina

The Human Change Management Body of Knowledge (HCMBOK®), Third Edition
Vicente Goncalves and Carla Campos

Creating a Greater Whole: A Project Manager's Guide to Becoming a Leader
Susan G. Schwartz

Project Management beyond Waterfall and Agile
Mounir Ajam

Implementing Project and Program Benefit Management

Kenn Dolan

CRC Press
Taylor & Francis Group
Boca Raton London New York

CRC Press is an imprint of the
Taylor & Francis Group, an **informa** business

AN AUERBACH BOOK

"Microsoft® Office" is a registered trademark of the Microsoft Corporation.

"MSP®" and "PRINCE2®" are registered trademarks of AXELOS Limited. All rights reserved.

"PMBOK® Guide" and "Pulse of the Profession®" are registered marks of the Project Management Institute, Inc., which is registered in the United States and other nations.

CRC Press
Taylor & Francis Group
6000 Broken Sound Parkway NW, Suite 300
Boca Raton, FL 33487-2742

First issued in paperback 2022

© 2019 by Taylor & Francis Group, LLC
CRC Press is an imprint of Taylor & Francis Group, an Informa business

No claim to original U.S. Government works

ISBN 13: 978-1-03-247591-2 (pbk)
ISBN 13: 978-1-4987-8639-3 (hbk)

DOI: 10.1201/9780429052972

This book contains information obtained from authentic and highly regarded sources. Reasonable efforts have been made to publish reliable data and information, but the author and publisher cannot assume responsibility for the validity of all materials or the consequences of their use. The authors and publishers have attempted to trace the copyright holders of all material reproduced in this publication and apologize to copyright holders if permission to publish in this form has not been obtained. If any copyright material has not been acknowledged please write and let us know so we may rectify in any future reprint.

Except as permitted under U.S. Copyright Law, no part of this book may be reprinted, reproduced, transmitted, or utilized in any form by any electronic, mechanical, or other means, now known or hereafter invented, including photocopying, microfilming, and recording, or in any information storage or retrieval system, without written permission from the publishers.

For permission to photocopy or use material electronically from this work, please access www.copyright.com (http://www.copyright.com/) or contact the Copyright Clearance Center, Inc. (CCC), 222 Rosewood Drive, Danvers, MA 01923, 978-750-8400. CCC is a not-for-profit organization that provides licenses and registration for a variety of users. For organizations that have been granted a photocopy license by the CCC, a separate system of payment has been arranged.

Trademark Notice: Product or corporate names may be trademarks or registered trademarks, and are used only for identification and explanation without intent to infringe.

Publisher's Note
The publisher has gone to great lengths to ensure the quality of this reprint but points out that some imperfections in the original copies may be apparent.

Visit the Taylor & Francis Web site at
http://www.taylorandfrancis.com

and the CRC Press Web site at
http://www.crcpress.com

Dedication

La Famiglia

Thank you for your love, support, and endless laughter, pride, and joy.

Jos, Ellis, Kyle, Andy, and Jackie

Roots and Wings

Margaret, Kenn, and Stuart

For your love,

And for teaching me the value of education and hard work.

Contents

Dedication	vii
Contents	ix
Preface	xvii
Acknowledgments	xix
About the Author	xxi
Introduction	xxiii

Part One
Establishing the Language — 1

Chapter 1: What Are Benefits and Why Are They Important? — 3
- 1.1 What Are Benefits? — 5
  - 1.1.1 Characteristics of Benefits — 6
  - 1.1.2 Why Are Benefits Important? — 8
- 1.2 Benefits Realization Management — 9
- 1.3 Categorization of Benefits — 11
  - 1.3.1 Financial and Nonfinancial Benefits — 13
  - 1.3.2 Triple Bottom Line — 15
  - 1.3.3 Efficiency and Effectiveness Factors — 16
  - 1.3.4 Tangible and Intangible Benefits — 18

1.4	Stakeholders and Benefits		20
1.5	Changing the Conversation		21
Exercises and Activities			23

Chapter 2: Definitions and Terminology — 25

2.1	Project		25
2.2	Program		26
2.3	Portfolio		27
2.4	Initiative		27
2.5	Progressive Elaboration		27
2.6	Pathway to Benefits		28
	2.6.1	Output	28
	2.6.2	Capability	30
	2.6.3	Outcome	31
	2.6.4	Intermediate Benefits	31
	2.6.5	Emergent Benefits	31
	2.6.6	Dis-Benefits	32
	2.6.7	Benefits	33
2.7	Transition Period		33
2.8	Transition Activities		33
2.9	Baseline		34
2.10	Operational Environment		34
2.11	Business as Usual		35
2.12	Gateways		35
2.13	Benefits Reviews		36
2.14	Fiscal Year (or Financial Year)		36
2.15	Front-End Loading		36
2.16	Return on Investment		36
2.17	Benefit Cost Analysis		37
	2.17.1	Payback Period	37
	2.17.2	Net Present Value	37
	2.17.3	Benefit–Cost Ratio	37
2.18	Diagrammatic Methods		38
	2.18.1	Benefit Map	38
	2.18.2	Dependency Network	38
2.19	Documentation		39
	2.19.1	Benefits Management Strategy	39
	2.19.2	Benefit Profile	39

	2.19.3	Benefits Register	40
	2.19.4	Business Case	40
	2.19.5	Benefit Realization Plan	40
	2.19.6	Program Plan	40
	2.19.7	Transition Plan	41
	2.19.8	Sustainment Plan	41
	2.19.9	Review Report	41
	2.19.10	Benefits Closure Report	42
2.20	Summary		42
	Exercises and Activities		42

Chapter 3: Team Roles and Responsibilities — 43

- 3.1 Sponsoring Group — 45
- 3.2 Sponsor — 48
- 3.3 Program Manager — 49
- 3.4 Project Manager — 50
- 3.5 Program/Project Management Office (PgMO/PMO) — 51
- 3.6 Project Office — 52
- 3.7 Program Office — 53
- 3.8 Business Change Manager (BCM) — 53
- 3.9 Change Team — 55
- 3.10 Benefit Manager/Owner — 56
- 3.11 Assurance Roles — 56
- 3.12 Governance Roles — 57
- 3.13 Specialist Support Roles — 58
- 3.14 Additional Considerations — 60
- 3.15 Summary — 60
- Exercises and Activities — 61

Part Two
The Benefits Life Cycle — 63

Chapter 4: Introduction to the Benefits Life Cycle — 65

Chapter 5: Establish the Context — 69

- 5.1 Drivers for Programs and Investments — 70
  - 5.1.1 PESTLE — 70
  - 5.1.2 Triple Bottom Line — 72

	5.2	Other Program Types	74
		5.2.1 Strategic Initiatives	75
		5.2.2 Evolving Initiatives	76
		5.2.3 Compliance Initiatives	77
		5.2.4 Technology-Driven Initiatives	79
	5.3	Recognizing the Stakeholders' Perspectives	79
		5.3.1 Getting to Know the Stakeholders	80
	5.4	Documentation	80
		5.4.1 Benefits Management Strategy	80
	5.5	Summary	81
		Exercises and Activities	83

Chapter 6: Identify the Benefits — 85

	6.1	Getting Off to a Bad Start . . .	86
	6.2	Begin with the End in Mind	87
	6.3	Diagrammatic Techniques	88
		6.3.1 Benefit Mapping	88
		6.3.2 Benefits Dependency Network	92
		6.3.3 Benefits Dependency Map	94
		6.3.4 Benefits Logic Map	95
		6.3.5 Applying These Methods	97
	6.4	Identifying the Right Benefits	98
	6.5	Who Identifies the Benefits?	100
	6.6	Documentation	102
		6.6.1 Benefit Profile	102
		6.6.2 Benefits Register	103
		6.6.3 Benefits Map (or Other Diagrammatic Representation)	105
	6.7	Summary	105
		Exercises and Activities	106

Chapter 7: Assess the Benefits — 109

	7.1	Quantifying	111
		7.1.1 Cognitive Bias—Some of the Traps	111
		7.1.2 Combatting the Biases	126
		7.1.3 Calculating the Value of Benefits	129
	7.2	Assessing	136
	7.3	Documentation	141

		7.3.1	Benefit Profile	141
		7.3.2	Benefits Realization Strategy	141
		7.3.3	Business Case (Initial)	141
		7.3.4	Benefits Register	142
	7.4	Reviewing and Decisions		142
		7.4.1	Initial Business Case Document	142
		7.4.2	Assurance	143
		7.4.3	Independent Review	144
		7.4.4	Learning Lessons	145
	7.5	Summary		147
	Exercises and Activities			148
Chapter 8: Plan for Benefits Realization				**149**
	8.1	Who Needs to Be Involved in Planning Benefits		151
	8.2	The Planning Regime for a Benefit Life Cycle		152
		8.2.1	Project Planning	156
		8.2.2	Transition Planning	158
		8.2.3	Post-Transition—After the Outcome Has Been Realized	162
	8.3	Planning for Benefits		163
		8.3.1	Early Wins	166
	8.4	Documentation		176
		8.4.1	Program Plan	176
		8.4.2	Benefit Realization Plan	177
		8.4.3	Transition Plan	178
		8.4.4	Sustainment Plan	178
	8.5	Summary		179
	Exercises and Activities			181
Chapter 9: Coordinate and Realize the Benefits				**183**
	9.1	Pre-Transition		185
		9.1.1	Changes to the Project	185
		9.1.2	Scope Changes	186
		9.1.3	Schedule Changes	186
		9.1.4	Training	187
		9.1.5	Communications	187
		9.1.6	Baseline	188
	9.2	Transition		190

		9.2.1	Induction	191
		9.2.2	Training	191
		9.2.3	Reinforcement	192
		9.2.4	Outcome	193
	9.3	Post-Transition		194
		9.3.1	Stepping Stones	194
		9.3.2	Minor Adjustments	195
		9.3.3	Reinforcement	196
		9.3.4	Measuring Benefits	196
		9.3.5	Decommissioning Obsolete Systems	197
	9.4	Sustainment		198
	9.5	A Case Study		199
		9.5.1	Pre-Transition	200
		9.5.2	Transition	202
		9.5.3	Post-Transition	202
	9.6	Documentation		204
	9.7	Summary		205
	Exercises and Activities			206

Chapter 10: Review the Initiative — 207

10.1	Addressing Failure to Meet Benefits Targets	211
10.2	Managing Emergent Benefits	217
10.3	Revisiting Planning	218
10.4	Closure	219
10.5	Documentation	221
	10.5.1 Review Report	221
	10.5.2 Benefits Closure Report	222
10.6	Summary	223
Exercises and Activities		223

Part Three
Embedding the Practices — 225

Chapter 11: Embedding Benefits Realization Management into Organizations — 227

11.1	Change the Conversation	228
11.2	Enforce the Development of Benefit Profiles	228

11.3	Apply Successful Delivery Mechanisms	229
11.4	Integrate BRM with Existing Organizational Processes	230
11.5	Induct All Stakeholders	230
11.6	Establish a Single Sponsoring Group	231
11.7	Focus on the Significant Benefits	232
11.8	Substantiate the Attribution of Benefits	234
11.9	Test the Legitimacy of Benefits	235
11.10	Beware "Double Dipping"	235
11.11	Apply a Model for Change	236
11.12	Be SMART	238
11.13	Engage Stakeholders	239
11.14	Conduct Independent Assurance and Reviews	240
11.15	Create Champions	240
11.16	Summary	241
	Exercises and Activities	242

Appendix I: Documentation — 243

Appendix II: Summary of Cognitive Biases Impacting Benefits Realization Management — 255

Abbreviations and Acronyms — 259

References — 261

Index — 265

Preface

Increasing accountability for expenditure and transparency to all stakeholders has led to benefits realization management being brought to the fore within the program and project management environments. It is, increasingly, being accepted as a function of the program manager, in particular, and a skillset with which the project manager must become comfortable.

I am passionate about benefits management because it elevates the project management profession and its professionals to a strategic level of engagement and contribution in the advancement toward the objectives. This involvement opens up new territory and raises problems and questions which may be new. Fundamental to these is WHY? Why is an investment being made? What do the investors and other stakeholders want from their investment?

Project management practitioners are expected to take on additional responsibilities and functions within their team. They are the inspiration behind this book. My aim was to write an interesting and supportive book which would allow practitioners to understand some of the complexities of benefits realization management with confidence. While theories are important, it is the application of them which adds value in the real world.

As someone who places great store on learning and education, I wanted to present the topic in a straightforward and practical way. This accounts for the anecdotes and case studies within the text. As the writing progressed, I enjoyed discovering and integrating onto the overall processes many of the "softer" aspect of benefits and their relevance to the program and project managers.

I hope that an initial reading will illuminate the subject but suggest that referring to the chapters and topics as they become relevant within the readers' work environment will enable the subject to come to life and apply the techniques.

Acknowledgments

As I have recently discovered, writing a book is not as simple nor as glamorous as it sounds. Many significant and unheralded contributions are needed order to complete the manuscript.

First, and most important, I thank my wife, Jo. She has been a large part of my professional and personal development for more than half my life. Her constant support has been a driving force in the development of this book.

Ellis and Kyle have been a constant source of inspiration. I continue to learn a huge amount from them and enjoy their support, humor, and conversation.

My mother, father, and brother provided the grounding in my formative years, and without their encouragement, wisdom, and guidance, none of this would have been possible.

I have been extremely lucky in my career to have met many generous professionals who have kindly given their time to listen, advise, and teach me—especially all of those "troublemakers" on training courses who have challenged me and forced me to consider and develop a greater understanding of the program and project world. There are too many to list, and I would not wish to miss any, so I hope you recognize yourself as you read this. I am extraordinarily grateful. This truly is a profession where its practitioners are willing to share and collaborate.

As a first-timer, I have been led through the process of writing and publishing by a strong and capable team. Thank you all!

Ginger Levin has been extremely encouraging and constructive in her advice and suggestions as the book evolved. Her patience and the words of wisdom of John Wyzalek (Taylor & Francis Group) have been essential to the completion of this project.

Theron Shreve, Lynne Lackenbach, and Marje Pollack (DerryField Publishing Services) have been invaluable assets and have patiently and gently nudged me along the pathway to publication. Their efforts have led to a better result.

Again, I thank you all!

About the Author

Kenn Dolan is a senior consultant and educator in program and project management. From a civil engineering background, he has managed a diverse range of projects and programs. As a consultant, Kenn has been engaged in several global transformation programs. He has focused on benefits realization management and stakeholder engagement since 2005.

With a Bachelor of Engineering degree from Imperial College, London, and a Master of Science degree from the University of Dundee, Scotland, Kenn is pursuing a doctorate through University College London.

Introduction

Increasing accountability for expenditure and transparency through governance arrangements and public access to information means that there is a greater need than ever before to ensure that projects are successful. To achieve this goal, it is necessary to understand the drivers for the investment and that expectations are clear and agreed. The project management profession must take the reins in these matters and ensure that the management of benefits is treated with the discipline it deserves and demands. The management of benefits will force the investors to be clear about the purpose of the investment and enable the team delivering programs, projects, and change to prioritize actions and make pertinent decisions.

The environment within which benefits are realized is complex. It includes the delivery of projects, changes within the operational environment, and the pursuit of strategic objectives and goals. Investors demand and expect returns on their investments; it is increasingly important that program and project managers can express the achievements of their efforts in concrete terms of relevance to the investors.

This is the language of benefits. It is a complicated and complex environment which must be understood and managed to avoid poor decisions being made or funds allocated unwisely. The link between the investment and the long-term benefits may be influenced by many factors. Rushing into a solution-focused mindset may provide results which are unexpected or unwanted. Lovins (1977) provided an excellent example of the consequences of failing to understand this environment which is paraphrased in the following:

> During the 1950s, malaria was rife among the Dayak people in Borneo. The World Health Organization (WHO) determined to resolve this major health issue by the use of large amounts of dichloro-diphenyltrichloroethane (DDT) as a powerful insecticide. This

succeeded in eradicating the mosquito population, as a result of which the incidence of malaria was reduced. All seemed good and the project was viewed as a success.

Shortly afterwards, the houses of the Dayak people suffered damage—the roofs collapsed. It was found that the use of DDT had killed the mosquitoes (as intended) but also the local wasps. These wasps fed on caterpillars, whose population surged in the absence of the predators. The caterpillars ate the thatched roofs, which promptly fell on the inhabitants. If that was not enough, the poison (DDT) found its way into the food chain, because the poisoned insects were eaten by the indigenous geckos and lizards. These were eaten by local cats. The cats died in large numbers and the rat population exploded. The rats brought disease, including typhus. Faced with another major health problem, the WHO took action by introducing 14,000 cats to address the rising vermin population. These cats were parachuted into Borneo.

Benefits realization management is not a linear process; by that I mean that one action does not always lead stepwise to the desired conclusion. Rather, programs, projects, and benefits are linked in a network in which actions can lead to unintended consequences or external factors can interfere with unexpected results.

This book aims to address the issues associated with realizing benefits. It is designed for program and project management practitioners to develop an appreciation for this value-adding skillset. Understanding the tools and techniques will offer an understanding of the quantitative aspects of discipline. However, this book will also address the mindset changes which need to be adopted.

This book is divided into three parts:

Part 1. Establishing the Language

The first three chapters focus on the need for a common understanding of terminology, to create a language which can be adopted to avoid ambiguity and misunderstanding. This part will also discuss the roles within a program environment and how they relate to benefits realization management.

Part 2. The Benefits Life Cycle

The benefits life cycle will be discussed in depth over seven chapters:

- Introduction to the Benefits Life Cycle
- Establish the Context
- Identify the Benefits
- Assess the Benefits

- Plan for Benefits Realization
- Coordinate and Realize the Benefits
- Review the Initiative

Part 3. Embedding the Practices

The final part, and chapter, is a discussion of the need to embed the practices of benefits realization management within an initiative or organization.

The processes are designed to provide guidance on the activities to be undertaken and the information to be gathered and deduced to make informed decisions. However, the practice of benefits realization management must take into account the culture and context within the organization.

Part One

Establishing the Language

Chapter 1

What Are Benefits and Why Are They Important?

> *"If you can't measure it, you can't improve it."*
> — Peter Drucker

> *"Success isn't about how much money you make, it's about the difference you make in people's lives."*
> — Michelle Obama

> *"We don't do projects just because we can. We do projects because they are supposed to add value to our organizations."*
> — Mark A. Langley

A global survey of over 500 executives found that only 61% of high-impact projects yielded their intended benefits. These projects were instrumental to implementing strategic objectives within the organizations (PMI, 2016a). Looking at this pessimistically, 39% of funded projects fail to deliver the benefits which are promised.

Additionally, projects continue to waste a significant amount of money. According to the Project Management Institute (PMI), US$99 million are wasted for every US$1 billion invested in projects (PMI, 2018). This is a significant improvement from previous results, but it still represents an enormous amount of poorly invested resources. The Global Infrastructure Hub (GI Hub; established by the G20*) forecasts that there is a need to spend US$94 trillion,

---

* The Group of Twenty (G20) is a forum for the governments and central bank governors from the largest global economies. It discusses policy about the promotion of international financial stability.

by 2040, to create the urban environments and critical infrastructure for the expanding global population (GI Hub, 2017). PricewaterhouseCoopers (PwC) estimates that annual global expenditures on capital projects and infrastructure will reach US$9 trillion by 2025 (PwC, 2014).

Organizations are committing to undertake programs and projects on an increasing scale as part of their corporate and public service manifestos. These programs and projects may be

- Infrastructure development
- Transformational change within an organization
- Technology and software development
- Research and development
- Oil-, gas-, or mining-related
- And many more from different organizations

Too often, the focus on projects has been to deliver the defined scope of the project, on time and within budget. Insufficient emphasis is placed on why the project is being undertaken and funded. Treating the project as an investment of time, energy, and funds forces the investors to adopt a different perspective of the project and the mechanisms for assessing its success. As with all investments, the investors expect a return. It is this return which will, ultimately, define the success of the project.

"You can't manage what you can't measure" is attributed to Peter Drucker. In a project context, this has traditionally led project teams to focus on the cost and time management of the project delivery, attempting to conform to the Triple Constraints of cost, time, and quality (or scope). However, these metrics apply only to the delivery of the project and do not tell the complete story. For example, delivering on time and within budget is good if the complete scope of the project is completed. Reducing the scope, or "de-scoping," to remain within the cost and time constraints will impact (negatively) the ability of the final product to enable the changes expected in the operational environment. Ultimately, this will lead to a reduction in the quantity of benefits which will be realized.

Drucker (2001, p. 28) stated: "The basic definitions of the business and of its purpose and mission, have to be translated into objectives. Otherwise, they remain insights, good intentions and brilliant epigrams that never become achievement." Drucker proceeded to detail the characteristics of these objectives:

- "Objectives must be derived from 'what our business is, what it will be and what it should be.' They are not abstractions."
- "Objectives must be operational. They must be capable of being converted into specific targets."
- "Objectives must make possible concentration of resources and efforts."

- "There must be multiple objectives rather than a single objective."
- "Objectives are needed in all areas on which the survival of the business depends."

So, from Drucker, to make change and transformation happen, we need:

- Measurable performance metrics which relate to the operational environment
- Criteria which are strategically important and related to the organization's raison d'être
- Objectives which can be converted into specific measurable targets
- Criteria which encourage and focus the application of resources that include funding

Failure to define the expected return and the specific benefits which can be measured, and are representative of the investment's objectives, will make it difficult to determine the true success of the program or project in objective and verifiable terms. With a clearly established metric for success, progress toward the goals can be monitored, and adjustments can be made to account for changing circumstances.

This introductory chapter will present the case for considering and applying benefits realization management (BRM) to the program and project environments. It will define and discuss the nature of "benefits" and how the focus and nature of programs, projects, and investments must change to create a healthier and more successful environment.

1.1 What Are Benefits?

A generic definition of the word "benefit" is that a benefit is something that is advantageous or good. In fact, the *Oxford English Dictionary* definition of "benefit," as a noun, is "A thing well done; a good or noble deed." In the management and corporate environments, the term "benefit" is used broadly and generically. When considering programs and projects, however, there is a need for a narrower, contextualized definition.

The Association for Project Management (APM) defines benefits as "[t]he quantifiable and measurable improvement resulting from completion of deliverables that is perceived as positive by a stakeholder" (APM, 2012). The Change Management Institute's (CMI) Body of Knowledge contains a similar perspective, defining benefits as "[t]he measurable improvement resulting from a change in the organization; and it offers an advantage to stakeholders who are inside or outside the organization." (CMI, 2013).

The PMI offers a value-based definition within its *Pulse of the Profession®* Report, which states that "project benefits are the value that is created for the project sponsor or beneficiary as a result of the successful completion of a project" (PMI, 2016b, p. 5), while *The Standard for Program Management—4th Edition* defines benefits as ". . . the gains and assets realized by the organization and other stakeholders as a result of outcomes delivered by the program" (PMI, 2017b, p. 44). And Jenner (2014) defines benefits as ". . . the measurable improvement from change which is perceived as positive by one or more stakeholders, and which contributes to the organizational (including strategic) objectives."

In summary, a benefit is measurable and cannot be referred to in vague terms—there must be a reliable way of measuring and assessing the value which results from the change. Additionally, the benefits identified as being the result of change should be linked to the organizational objectives. In practice, this may not always be the case. Benefits do not always have to be connected directly to the strategic and organizational goals and objectives. However, if an initiative is commissioned to deliver benefits which are *not* related to strategic and organizational objectives, a number of questions should be raised. The first among these questions is "Why?"

- Why are resources being deployed to achieve goals which are not recognized as being important strategically?
- Why are the resources not being directed toward initiatives which are connected to organizational goals?
- Why are decisions being made which are contrary to the objectives the organization is moving toward?

Oftentimes, initiatives are justified through a business case. Many business cases contain some benefits which are linked to organizational and strategic goals, *and* other benefits, which may be more tactical in nature. This balance is often necessary to gain the support of key stakeholders and balance long- and short-term priorities and desires.

Generally, there is agreement within the project management community that benefits represent measurable improvements, which have arisen through the delivery of a project(s) and some associated changes within the operational environment.

1.1.1 Characteristics of Benefits

Benefits should conform to a set of criteria before they can be claimed. *Managing Successful Programmes (MSP®)* (AXELOS, 2011) established criteria which must be applied to all benefits. Each benefit must satisfy the following criteria:

- Description
- Observable
- Attributable
- Measurable

In addition, Jenner (2014) advocates the technique of "booking the benefits." Based on the assertion of Kaplan (2005) that benefits should be built into financial forecasts, this technique dictates that a benefit cannot be claimed unless it is realized and allocated to future forecasts and operating budgets. For example, when a project is implemented to automate a process and will reduce the operating costs associated with that process because fewer operatives will be required, the benefit can only be claimed to have been realized once the operating costs have been reduced, the operatives made redundant, and the future operating forecasts adjusted.

In one instance, an organization automated and changed its financial management processes, which led to greater efficiency in the management of transactions to such an extent that when one of the members of the financial management team retired, there was no need to replace that role. This benefit was booked into the budget for the next fiscal year, when the business unit's operating budget was reduced. Had a replacement been appointed in a part-time or different capacity, the value of the benefit would have been reduced to take into account the actual reduction against the baseline.

I believe that this approach should be highlighted and introduce a fifth criterion: legitimacy. This makes the criteria for a benefit:

- Description—A clear description of the benefit must be produced. This will ensure that there is no confusion or misunderstanding about the benefit. This common understanding and agreement among the stakeholders will support the selection of the programs and the projects which will be implemented to enact the changes required to realize the benefits.
- Observable—A baseline must be established which can be used as a comparison for future-state performance levels. This is closely aligned to the "measurable" criterion. Choosing the appropriate baseline for analysis is an important factor in being able to point to a visible and verifiable difference between the "before" and "after" states of the change.
- Attributable—Benefits must be attributable to the change made. A strict enforcement of this criterion will drive an open and honest view of the achievements of the program and its associated change. Benefits can only be claimed if they are directly the result of actions taken within the

initiative. In other words, benefits cannot be recorded against the initiative if they are the result of circumstances and events outside the initiative; for example, a change in currency exchange rates makes the project more profitable than predicted.
- Measurable—A method must be established for measuring the benefit. This will force stakeholders to talk differently about the purpose of the change and its objectives. The use of vague terms such as "better" and "improved" will be removed from the vocabulary because they are open to subjective interpretation.
- Legitimacy—The benefits are realized as a result of the initiative and manifest themselves in forecasts and future budgets, or other performance measures. The benefits are genuine, and the changes are advantageous to the stakeholders.

1.1.2 Why Are Benefits Important?

Change initiatives, programs, and projects should be treated as investments. Regardless of their scope and initial purpose, someone, or often a group, commits funds and resources to undertake the work. The work is delivered in the belief that it is necessary to achieve a defined goal, which is often an improvement in performance that is valued more highly than the cost and effort of undertaking the initiative.

According to Jenner (2012), organizations struggle to demonstrate a return on their change-related investments, citing the Office of Government Commerce (OGC) (2003), which stated that "[d]eficiencies in benefits capture bedevils nearly 50% of government projects." Looking beyond the public sector, Lovallo and Kahneman (2003) found that most large capital investments are delivered beyond the scheduled completion and over budget—the investments fail to live up to the expectations of the key stakeholders.

Essentially, there is an expectation that the investment will provide a return. Benefits can be used as the measure of that return and to ensure that the focus of the initiative remains of the generation of value.

As an organization becomes more mature (with respect to the management of portfolios, programs, projects, and change), the decision to commission one initiative, rather than another, becomes more important. Indeed, this becomes a conscious decision made by executives to allocate resources in one way. Those which return the greater volume or, more important, strategic benefits are likely to gain greater support from stakeholders and executives in preference to others. When resources and funding are limited, understanding the benefits will be used to answer the question, "Where should the money be invested?"

It is regularly reported that 70% of change initiatives fail to deliver the benefits they were established to deliver. It has often been reported that success rates of change programs represented an unacceptable return on investment. One of the common reasons for these failures is a lack of focus on the real purpose of the initiative: to generate benefits. Particularly with projects, it is all too easy to become distracted by the technical objectives and outputs, which are created in the relatively short term, leaving the longer-term benefits overlooked.

Cooke-Davies (2002) proposed that there should be an important distinction between project management success and project success. Project management success includes considerations such as:

- Delivery of the project within the Triple Constraints:
  - Delivery on time
  - Delivery within budget
  - Delivery of the complete scope of the work
- Compliance with the organizational processes
- Useful and complete documentation
- Stakeholder engagement during delivery

These are important elements of the project whose importance should not be underplayed or overlooked. However, they make no contribution toward confirming whether the project achieved its goals or met the agreed-upon purpose. Was the investment successful? This can only be determined by identifying the benefits expected from the initiative and measuring them when they are realized. The success of the project should be defined in the same manner as any other investment—by comparing the resources expended to the benefits realized within the changed operating environment.

1.2 Benefits Realization Management

The APM Benefits Management Special Interest Group (2011) states, "[i]f value is to be created and sustained, benefits need to be actively managed through the whole investment lifecycle. From describing and selecting the investment, through program scoping and design, delivery of the program to create the capability and execution of the business changes required to utilize that capability, and the operation and eventual retirement of the resulting assets. Unfortunately, this is rarely the case."

Benefits management is defined as the identification, definition, planning, tracking, and realization of business benefits (APM, 2011). Jenner (2014) defines benefits management as "the identification, quantification, analysis, planning, tracking, realization and optimization of benefits." The terms "benefits management" and "benefits realization management" are often used interchangeably; "benefits realization management" (BRM) will be used throughout this book, simply because the realization of the benefits in the post-project period is extremely important.

Of the two definitions above, the second (Jenner, 2014) is the more complete and the more proactive in that it stresses the significance of the analysis and optimization of benefits. The process requires a constant, or at least a regular, review of progress toward the goals and revision of the plans and actions throughout the benefits life cycle. BRM should be viewed as a proactive process rather than a rigid and unyielding approach.

There should be clarity and consistent use of terminology, which will support the communications between the delivery teams and the operating uses and other stakeholders.

A project does not deliver benefits directly, as indicated in Figure 1.1; rather, they are the starting point of a journey to reach those longer-term goals. Projects are, by definition, designed to deliver a specified deliverable (output) or service.

Consider a project to construct a new bridge between two islands, designed to connect the main towns on each island. The project manager's role is to manage the delivery of the specified output—in this case, a bridge. The ownership of the output is then transferred to the client or operational group of users—for example, the Department of Transportation. The output must have been designed to be of use and value to the users or sponsoring organization; that is, it must be capable of operating in the new environment. In this example, the bridge will be designed to support specific traffic loads and numbers of vehicles.

The capability is the point at which the completed works can be put into operations and may include transitional tasks beyond the project to ensure that the step into use is smooth. Once the output is complete, it may need to be

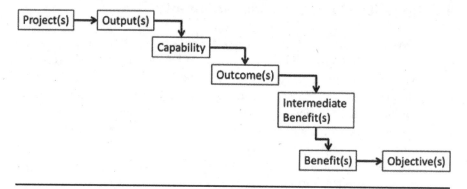

Figure 1.1 The Journey from the Project to Benefits and Objectives

combined with other outputs before it is ready for operations—for example, signage may need to be installed on existing roads, the toll booths may require the installation of a payment system, and toll booth operators need to be employed and trained, or perhaps an electronic toll system needs to be designed and installed. The capability is available when all of these outputs are combined into a system which can be put into use.

As a result, the outcomes will be realized when the output is used in the operational environment. The outcome may be described in terms of the operational state—for example, the bridge is in use. However, this outcome may be affected by other projects, such as the upgrade and construction of access roads from the towns to the site of the bridge, or the extension of highways to the bridge. As such, just because the project has been completed, there is no guarantee that the outcome will automatically follow.

Outcomes may lag behind the project, particularly if there are dependencies between one project and the completion of other projects, parts of a program, or even events beyond the control of the sponsoring organization.

Benefits, among other criteria, must be measurable. In the case of the bridge, the benefits may include:

- Reduced time of travel between the towns, compared to the previous methods of transport, which may have been by ferry or airplane
- Reduced cost of transportation of freight
- Number of vehicles using the new bridge
- Increased tourism, measured by the number of tourists visiting or revenue generated from tourist activities
- Revenue generated through the gathering of tolls for use of the bridge

Some of these benefits will be realized more quickly than others. Indeed, these earlier returns may be considered as "intermediate benefits," which are important indicators that future benefits are likely to be realized.

Ultimately, the full suite of benefits will be realized, and they should be representative of strategic and corporate objectives. Table 1.1 summarizes some examples of projects and their outputs and the journey which may be followed to realize the benefits.

1.3 Categorization of Benefits

Although each organization may wish to define benefits in a way that is consistent with its own culture and environment, there are some common categories which can be used when defining benefits:

Table 1.1 Examples of Outputs Leading to Benefits

Project	Output	Capability	Outcome	Benefits
e-Commerce	An element of an e-commerce system—such as hardware, software, and business processes	The commissioned system completed, with trained staff and support arrangements ready to go live	Customers using the new e-commerce platform and transactions being processed	• Faster processing of transactions • Increased number of transactions conducted electronically • Increased number of transactions overall • Lower costs of sales • Increased sales income • Increased profit
Airline	New aircraft purchased (with greater capacity and fuel economy)	• Aircraft delivered and sufficient air crew (pilots and cabin crew) trained • Ground crew and partners advised • Maintenance arrangements in place	Aircraft in service and operating to the performance metrics	• More passengers per flight • Reduced cost per passenger mile traveled • Increased customer satisfaction • Reduction in operating costs • Reduction in fuel costs • Increased profits
Hospital	New hospital building, including new equipment and electronic systems	• Commissioned building and equipment with trained medical and administrative staff	Hospital in use, with inflow of patients and procedures and services provided	• Reduced number of days patients remain in hospital • Lower costs of procedures • Reduction in waiting lists for specific procedures

- Financial and nonfinancial benefits
- Efficiency and Effectiveness
- Tangible and Intangible

Categorizing the benefits in this manner forces the delivery team and the stakeholders to view the initiative from a different perspective. Expectations change from the focus on the product to ensuring that the outputs lead to changes in the operational environment and among the users.

1.3.1 Financial and Nonfinancial Benefits

The question often arises, "Should all benefits be measured in dollars?"

The answer is "No!"

While all benefits must be measurable, that does not necessarily mean that they must be financial—although it does help if they can be expressed as a dollar figure. Expressing the benefits in financial terms makes the comparison of cost and benefit clear and unambiguous.

Consider a project investment of $10 million, which results in changes to the work environment generating $14 million in cost savings over the next three years. In this case the decision is relatively simple: The resulting benefits are clearly greater than the investment.

Would the decision be quite as simple if the proposal were for an investment of $1 million to undertake a project aimed at increasing staff morale?

First, it is important to stress that staff morale is important. However, it is difficult to make the decision to invest a large sum of money to gain a return which is not easily measured and is not financial. A leap of faith would be required in considering whether the investment is worthwhile and the project viable.

Second, there may be a way of changing the perspective to make the decision more straightforward. This can be achieved, in part, by the use of proxy measures. Instead of measuring staff morale by using staff satisfaction surveys, consider a more tangible metric which related to staff morale.

Consider the same proposal: $1 million investment resulting in improved staff morale. Staff morale can be measured through the use of surveys, questionnaires, and interviews, but it remains difficult to quantify the value of any change in this metric. However, improving staff morale leads directly to increased productivity and reduced turnover in personnel (fewer people leave the organization and have to be replaced). Assume that a more content team will remain with the organization and reduce the number of people leaving from 75 to 45 per year. This would save the organization the costs and effort associated with recruitment 30 times each year.

These costs include:

- Loss of production and productivity from the team, which has lost a member
- Payouts for holiday pay and personal leave accrued
- Advertising the position
- Reviewing the responses for the position, which is time-consuming
- Perhaps engaging a recruitment consultant—which may be costly
- Preparing a shortlist for interviews
- Conducting interviews, including loss of productive time from the team so they can be involved in interviews and assessments
- Induction costs for the successful applicant
- "Loss" of production and productivity as the team is disrupted by the change
- Short-term underperformance of the new recruit—there is likely to be a period of time during which the newly employed team member will not perform at the same level as the other team members

Records may indicate the costs associated with all these activities to be $10,000. The total saving per year would be $300,000. Suddenly the $1 million investment is considerably more attractive because of the link between improved morale and operating costs associated with the departure and recruitment of personnel. The project costs will be recovered within four years. The investment decision can be based on much more robust data because the proxy measures were applied.

Incidentally, the estimated costs associated with replacing an employee vary depending on the employee's role and specialist skills or knowledge. Boushey and Glynn (2012) found that this ranged from 16% of annual salary for employees earning less than $30,000, while replacing mid-level managers cost approximately 20% of their salary, and replacing high-valued professionals could cost the organization over 200% of the salary. Other sources estimate the cost of replacing an employee is 50–75% of his or her annual salary. Furthermore, these estimates account only for the actual costs associated with a member of the organization resigning and the recruitment of a replacement—they do not reflect the value of the loss of corporate knowledge and wisdom.

Benefits may be described in financial (sometimes referred to cashable) terms, or in nonfinancial (sometimes referred to noncashable) terms. Some examples of nonfinancial benefits are

- Reduction in carbon dioxide (CO_2) emissions—Emitting lower quantities of damaging chemicals and gases during a particular manufacturing process.
- Reduced volume of waste—Reducing the volume of waste from a process may make that process more efficient and ultimately reduce costs. Less

effort will be required to recover materials and dispose of them—which may be a significant cost.
- Reduced travel time—Reducing the amount of time spent traveling to and from the workplace would be an advantage to employees. Also, reducing the time spent traveling to meetings, for example, by implementing an effective video conferencing system, would reduce traveling costs and increase the time available for other activities.
- Increased time available for other activities—Reducing the time spent on some activities, or eliminating those activities altogether, would allow time to be devoted to other tasks.
- Number of steps in process—Simplifying a process is often valuable when a key component of success is to encourage large numbers of users to use the new system. A more streamlined process, with fewer steps, is less likely to discourage users from signing up and beginning to use the new system.
- Increased number of users—Increasing the number of people using a system may be the primary purpose of a project or initiative. Increasing the number of users on one system may result in the older system becoming obsolete or may reduce the operational costs of the new system.

As can be seen from the examples above, some of the nonfinancial benefits may be linked to financial benefits which follow later. However, there may be a lag between the achievement of a nonfinancial benefit and the subsequent financial ones, so it is important, when promoting a new initiative, that the focus is maintained on elements which are valued by the stakeholders. For example, reconfiguring a public transportation system (or at least part of the overall system) may result in reducing travel times and improving reliability. This may lead to an increase in passengers because the system is more reliable (and a financial benefit of increase revenue) and a reduction in the operating cost per traveler—another financial benefit. Although there may be an ultimate financial consideration, recording the nonfinancial benefits will ensure that they remain indicators of the likelihood of generating the longer-term benefits.

1.3.2 Triple Bottom Line

Many organizations adopt a holistic approach to their investment decisions and performance reporting by applying the Triple Bottom Line (TBL) framework. The term Triple Bottom Line was first used by Elkington in 1994, and detailed by Elkington (1997), and refers to the measurement of an organization's performance not merely based on profit or loss criteria, but using a broader range of metrics:

- Economic factors
- Societal factors
- Environmental factors

This allows organizations in the public and private sectors to assess their performance taking into account a range of considerations linked to their interpretation of their corporate social responsibilities (CSR). Applying the Triple Bottom Line framework enables organizations to weight (not necessarily equally) nonfinancial factors as part of their performance reporting, for example, the creation of local, long-term jobs.

Organizations may make large profits through their manufacturing processes, but the waste products may cause damage to the environment and illness among the local population. Government bodies then need to spend significant resources to clean up the environment and manage the health issues of those affected. Although there may be a profit for the manufacturing organization, there may be a net negative impact on the community and the local economy. The application of the TBL records the performance of all three facets of the organization's activities and provides a balance between the stakeholders, being shareholders and owners, the local community, and the wider environment, offering a more complete perspective of the impact of the organization's activities.

In the context of benefits realization management, this encourages the use of nonfinancial benefits within the business case. For example, the following nonfinancial benefits may be included:

- Social benefits
  - Creation of jobs—This is not a benefit to the organization sponsoring the initiative, but it is a benefit for the community.
- Environmental benefits
  - Reduction of waste materials produced
  - Increase in water quality

Depending on the circumstances and organizational philosophy, these nonfinancial benefits may be included in a business case and weighted against the financial benefits to enable a broader comparison of the investment and its total value.

1.3.3 Efficiency and Effectiveness Factors

Understanding the intention of the initiative or program and its intended impact will assist in the identification of specific and valued benefits. Categorizing benefits as being directed at efficiency or effectiveness gains will clarify the intent of the changes and/or results.

Efficiency describes the balance between the resources required to deliver the changes created. It is a comparison between inputs and outputs. Examples of benefits which fit the efficiency category are

- Improvements in productivity, which is calculated by dividing the value created by the resources used as inputs into the project and subsequent changes. This can be achieved by increasing the value created with the same resources or producing the same value with fewer resources required.
- Reduction in the number of steps in a process—requiring fewer inputs to achieve the same results.
- Simplification of business processes, allowing the redeployment of some resources to undertake value-adding work. In essence, this enables the creation of additional value from the same resources.

Effectiveness reflects the degree to which actual results align with the intentions of the initiative. It can also relate to the extent to which the initial problems are solved. Effectiveness benefits include:

- Improved governance of an organization, exemplified by more accurate or timely performance reporting.
- Improved operational processes creating a higher-quality product. This could be measured by a reduction in test failures.
- Improved customer service as measured by the satisfaction of customers through the telephone support function or measured as a percentage of calls which result in a successful outcome for the customer.

Effectiveness benefits are not usually described in financial terms, although they may ultimately lead to financial benefits, for example:

- A higher quality product—the benefit is measured as the reduction in test failures
  LEADS TO
- A reduction in overall production costs—which is a financial benefit
  AND
- An improvement in the perception of the product as a quality product
  LEADS TO
- Increased sales

In summary:

- *Efficiency* is about doing things right—improving productivity, reducing costs and waste.
- *Effectiveness* is about doing the right things—linking benefits to organizational strategy and making better decisions.

1.3.4 Tangible and Intangible Benefits

Tangible benefits are those which are physical in nature or easily measured, for example:

- Quicker processing time
- Reduction in effort required
- Productivity improvements
- Reduction in travel time

These benefits cannot be applied to all initiatives. However, they have the significant advantages of visibility to stakeholders and users, and they can be measured in a relatively straightforward manner. These benefits have a clear and unambiguous meaning to the stakeholders.

Some benefits are not physical, or easy to measure, and these are often referred to as intangible benefits. Examples of intangible benefits are:

- Safer community
- Brand recognition
- Improved quality of life
- More user-friendly application
- "Greener" outcome

It is tempting to introduce these, often nebulous, factors into the business case and including them in the investment decision because they can be used to build convincing arguments for proceeding. However, the intangibility of these benefits should be a warning to sponsors because they are, by their nature, difficult to measure and open to interpretation, and therefore, the overall success can be debated. The initiative may proceed with few defined success criteria (benefits) and with no meaningful manner by which the business case will be evaluated for investment, progress, or success. There is also the problem of misinterpretation of the intangibles by the stakeholders—that is, each stakeholder may have a different understanding of the benefit.

What Are Benefits and Why Are They Important? 19

This may lead to differing expectations among the stakeholders and problems with misunderstandings of communications.

For example, if the aim of a local government initiative is to create a "safer community," how will that be measured? Various stakeholders may perceive the objectives differently and expect improvements such as:

- Reduction in crime
- Increase in arrests for criminal offences
- Increase in conviction rate for offenders
- Increased lighting in public places—streets and parks
- Increased number of CCTV cameras in a retail area
- Support for the installation of security measures such as monitored alarms and grills for windows

It is often helpful to devote time and resources to the definition of the intangible benefits to create a common understanding among the stakeholders regarding the nature of the benefits. This may result in one of two conclusions:

- Some of the intangible benefits can, with research, be converted to, or defined as, a tangible metric.
- Some of the intangible benefits cannot be measure directly but can be measured through a proxy measure. A proxy metric is one which is related to the intangible benefit but does not measure that benefit directly. The proxy metric is a legitimate indicator that the intangible benefit exists. The "increased user-friendliness" of an application is difficult to define. However, it is likely to lead to an increase in the number of users taking up the application. This will generate revenue or drive cost savings to the operations of the organization through a reduction in transaction costs. User numbers, revenue, and transaction costs can be measured as tangible results, which are related to objectives.

There are benefits which are not tangible. These may be recorded in the business case. Many useful business cases contain a mixture of tangible and intangible benefits. However, building a business case which is based solely or predominantly on intangible benefits will create problems when attempting to justify the investment. Considering the measurement of benefits in future years and reflecting on the success (or perceived success) of the initiative, audits are not always sympathetic to the recording of intangible benefits. Table 1.2 summarizes the different categories of benefits and offers some examples which may be useful in establishing the relevant mindset among the stakeholders.

Table 1.2 Summary of Benefit Categories

Category	Definition	Examples
Financial	A benefit which can be measured in monetary terms.	Reduced cost of transaction Increased revenue or profit Reduction in operational costs
Nonfinancial	A benefit which is not measured in monetary terms.	Reduction in equipment downtime Increased reliability of service Reduced number of steps in a process
Efficiency	Efficiency describes the balance between the resources required to deliver the changes created. It is a comparison between inputs and outputs.	Increased productivity Reduced cost of service Reduced number of steps in a process
Effectiveness	Effectiveness reflects the degree to which actual results align with the intentions of the initiative.	Improved governance Enhanced product features Availability of timely reports
Tangible	Tangible benefits are those which are physical in nature or easily measured.	Quicker processing time Productivity improvements Reduction in travel time
Intangible	Intangible benefits are difficult or impossible to measure reliably.	Safer community Brand recognition Improved quality of life

1.4 Stakeholders and Benefits

Stakeholders are likely to have different perspectives of what constitute benefits. They will certainly have differing levels of interest in the benefits, which are recorded in the business case. Schueler, Stanwick, and Loveder (2017) identified three different types of returns as being expected from the delivery of training and education initiatives. Although this work focused on the higher and vocational education and training sectors, it is a strong reminder that stakeholders have expectations that the returns will be relevant to them individually as well as through the sponsoring organization. Benefits are more likely to be perceived as more important if they serve stakeholders' interests. The three types of returns can be summarized as:

- Personal—returns which impact the individual or employee
- Organizational—returns which address the needs of the organization(s) sponsoring the initiative
- Community—returns which address the local, regional, or national economy and residents

The European Centre for the Development of Vocational Training (CEDEFOP) referred to these three levels (CEDEFOP, 2011) as:

- Micro—relating to the individual
- Meso–relating to the organization
- Macro—relating to the community

The categorization of the benefits by the stakeholder groups, as in Figure 1.2, will clarify the intent of the initiative and its impact on different groups within the community. It will assist with the communications between the initiative and the stakeholders by focusing on the issues and returns which impact the stakeholders directly and which they have an interest in pursuing.

1.5 Changing the Conversation

Programs and projects need significant funding either from the shareholders of the sponsoring organization or from taxpayers for government-funded initiatives. Societal awareness and maturity demands greater accountability of the funding and objectives of these initiatives and their need and objectives.

One of the most important aspects of benefits realization management is awareness of the purpose of the initiative (whether it is a through a project or a program) and the need to think beyond the outputs which are created in the process.

In many ways, the widely accepted definition of a project has held back benefits realization management. Defining the objective of a project as being the delivery of a product or service means that the focus of all of the effort is on that product. Once the product is created, the project ends. Success is defined by the delivery process alone.

Focusing on the aftermath of the project forces a more disciplined approach to justifying the investment. The implementation of benefits management will lead to a number of changes in the psyche of the organization and teams. Team members, stakeholders, and management will begin to view projects differently and discuss them in a new light. Projects will no longer be seen as burdens on the organization but as opportunities for investment and change. The investment and portfolio environment will be seen in a more complex and accurate way. The misperception that the delivery of a project in isolation will lead automatically to the generation of benefits in a linear fashion will be discarded. The links between projects and benefits are much more complex and more similar to a network, in which several projects may be connected to the same benefit. To be successful in this environment, a great deal of coordination is required

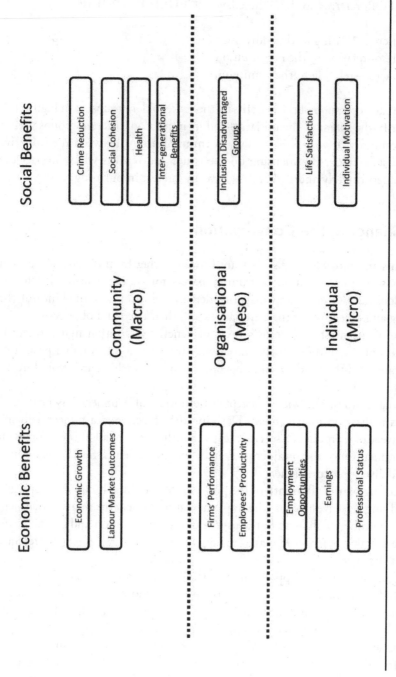

Figure 1.2 The Benefits of Vocational Education and Training (Modified from CEDEFOP, 2011)

between the projects and transitional activities. Initiatives are more likely to achieve their objectives if they are managed as part of a portfolio or program-based approach.

Most important, the discussion regarding projects will change from a product-oriented one—that is, what things the project will create—to a results-based one—what opportunity we will gain by delivering the project and what advantages it will enable in our operational environment. This change in the discussion is important because it creates an environment in which decisions will be approached differently; they will be based on business needs and results rather than be technically driven. The selection and prioritization of projects will change, and the allocation of resources within the organization will take on a different perspective. Projects which deliver the greatest return on investment will be prioritized over less attractive investments.

From a project manager's perspective, it will become easier to engage with key stakeholders and to obtain enthusiastic buy-in from business units. Projects and changes will be designed to create measurable improvements and advantages which are important to those business units. This should not be seen as merely a better way of marketing and promoting the projects. Changing the conversation will enable all parties to express their needs in terms which are unambiguous and clear. There will be a common understanding of the intent and needs of the projects, which will encourage and require the stakeholders to involve themselves more within the project, and at an early stage. The position of the Program Management Office (PgMO) or Project Management Office (PMO) is often strengthened because of the pivotal role it can play in the deployment of benefits realization strategies within the organization.

This is not a wish list! These observations and conclusions are derived from many engagements in consulting roles within a number of organizations over a lengthy period of time. Changing the conversation about projects will result in a stronger stakeholder engagement and better decisions being made about projects, especially their selection, justification, and management of their delivery.

The next chapter will define and discuss the terminology which will be used throughout this text. It is important that the language is well defined in order to change the conversation. Additionally, the next chapter will discuss the documentation which will be useful and the purpose of each document.

Exercises and Activities

1. Discuss which categories of benefits are appropriate in your experience and your organization's environment.

2. Using the CEDEFOP model, identify benefits which would fall into each of the three categories:
   a. Micro
   b. Meso
   c. Macro

Chapter 2

Definitions and Terminology

"The beginning of wisdom is the definition of terms."
— Socrates

"The enemy of accountability is ambiguity."
— Patrick Lencioni

"England and America are two countries divided by a common language."
— George Bernard Shaw

The use of terminology must be clear and consistent when changing the conversation. The use of terminology becomes important when discussing benefits realization management (BRM) with stakeholders who may not be as familiar with change initiatives as the delivery team. To ensure clarity of message, the language used should be unambiguous and consistent. There should be an accepted explanation for all common terms and a point of reference for all stakeholders to support all of the discussions relating to the initiatives and their benefits.

This chapter defines the use of the common terms which will be used throughout this book.

2.1 Project

The *PMBOK® Guide*, 6th Edition, defines a project as ". . . a temporary endeavor undertaken to create a unique product, service, or result" (PMI, 2017a).

PRINCE2® defines a project as ". . . a temporary organization that is created for the purpose of delivering one or more business products according to an agreed business case" (AXELOS, 2017).

The PRINCE2® definition directly relates the need for a project to connect with the "business" objectives of the sponsoring organization. This reinforces the view that projects should be considered as investments and that there should be a valid business case established to justify the funding of the project. This method of justification will include an investment appraisal, which will compare the scale of the investment with the rewards that will be realized and measured as benefits.

The major issue with both definitions is that although benefits realization management is inferred through the business case, the focus of the project is the delivery of the unique product. Benefits are predominantly realized following the closure of the project. This often creates a disconnect between the delivery team and the operational owners who will take possession of the products.

The two definitions cited above are similar, and other definitions could also be used. For the purposes of this book, the PRINCE2® definition will be used because of the explicit reference to the business case.

2.2 Program

The Standard for Program Management, 4th Edition, defines a program as ". . . related projects, subsidiary programs, and program activities managed in a coordinated manner to obtain benefits not available from managing them individually" (PMI, 2017b). In *Managing Successful Programmes (MSP®)* (AXELOS, 2011), the definition of a program is ". . . a temporary, flexible organization created to coordinate, direct and oversee the implementation of a set of related projects and activities in order to deliver outcomes and benefits related to the organization's strategic objectives."

An important common feature of both definitions is that a program consists of projects *and* other activities which are necessary to the ultimate realization of benefits. These other activities are undertaken outside of any of the projects and include:

- Coordination activities
- Governance and assurance activities
- Transition from the project to an operational environment
- Benefits measurement

The definition from *The Standard for Program Management* (PMI, 2017b) will be used throughout this text, because it refers to the establishment of a program to achieve advantages which are not available by managing the projects separately. Since it is not always necessary to establish a program, a major consideration should be whether there are advantages to creating a centralized, overarching coordination mechanism as opposed to managing the projects separately.

2.3 Portfolio

The Standard for Portfolio Management, 4th Edition, defines a portfolio as a "... collection of projects, programs and subsidiary portfolios, and operations managed as a group to achieve strategic objectives" (PMI, 2017c). In the Office of Government Commerce (OGC)'s *Management of Portfolios*, a portfolio is defined as "An organization's change portfolio is the totality of its investment (or segment thereof) in the changes required to achieve its strategic objectives" (OGC, 2011).

In this text, "portfolio" will follow the OGC definition because it specifically addresses the need to consider the total investment being made within the organization. The "totality of its investment" is a reference to the enterprise-wide investment in and commitment to projects, programs, transitional activities, and the operational environment.

2.4 Initiative

"Initiative" is a general term used to refer to an investment, program or project. It is used in this book to avoid repetition of the term "program and project"; whenever the term is used, it indicates a portfolio, a program, or a project, or a combination of these. The term is often used within organizations to indicate the activities associated with a change or the implementation of strategy.

"Initiative" is often used in the context of change. Change initiatives are similar to programs and may be managed as such. A change initiative may consist of several distinct packages of work plus some coordinating activities and tasks associated with embedding the change within the operational environment.

2.5 Progressive Elaboration

Programs and projects are complex undertakings. Often, not all elements are clearly understood at the commencement of the initiative, and a greater understanding of its scope, content, and impact will be developed as the work progresses. This increased understanding and more accurate information will influence future decisions and changes to elements of the initiative or even its overall direction.

"Progressive elaboration" is a legitimate approach to complex undertakings, which allows, and encourages, controlled changes to be made as additional, more accurate and relevant information comes to light through the delivery process. This approach may be applied to both programs and projects, although change needs to be controlled, approved, and managed through the appropriate mechanisms. It could be argued that progressive elaboration is more appropriate to the program environment, which is by nature more flexible, and because of the longer duration of a program, as it is subject to more external influences.

2.6 Pathway to Benefits

Figure 2.1 shows a more complete flow through the benefits life cycle from action within the projects to the strategic objectives. In summary,

- Projects create Outputs through the use of resources and funding.
- Outputs should be coordinated with changes within the workplace, or operational environment, to establish the Capability, before the Transition Period can commence. These changes may include:
  - Training of personnel
  - Development of new processes or activities in the operations
  - Recruitment of new skillsets
- With the Capability in place, the transition to the new operating model may commence. At the end of the transition period, the Outcome will be created and should be sustainable. The Outcome represents the new operational state—the Capability is being applied to the new operating environment (the *new* "business as usual").
- Following the achievement of the Outcome, the short-term value of the change may be measured through the Intermediate Benefits, and it is possible that new benefits are identified. These Emergent Benefits are unanticipated advantages, which are noticed or forecast as the initiative progresses and a greater understanding of the changes is acquired. There may also be negative consequences as a result of the changes, which need to be accounted for—these are known as Dis-Benefits.
- In the longer-term, the final Benefits will be realized and measured, and these should be linked to, or represent, the organization's Strategic Objectives.

2.6.1 Output

An output is the result of effort conducted within a program or project. The terms "product" and "deliverable" are used interchangeably with "output"; there

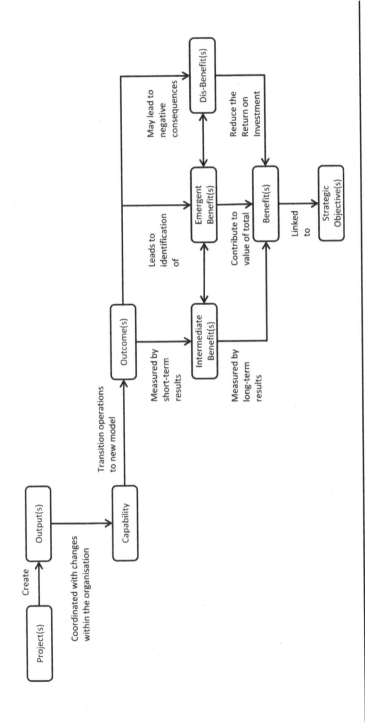

Figure 2.1 Detailed Pathway of the Development of Benefits

is often a preference to use one of these terms based on industry, profession, and culture. There are two types of products within the project environment.

- Management products
  These are primarily documents which are used by the delivery teams within the initiative, including:
  o Plans
  o Progress reports
  o Risk and issue management documentation
  o Documentation to track and control changes within the program
  These documents are not normally handed over to the sponsoring groups at the completion of the initiative. They are a means of managing the work of the project and making decisions within the project environment.
- Technical products
  These are the outputs, which are completed by the delivery team whose ownership is transferred to the operational teams. These products include:
  o Software and hardware
  o New and refurbished facilities and equipment
  o Training and induction assets
  o User guides and operating manuals
  o Reports and presentations of findings of an investigation
  o Recommendations from a review

The term "product" will be used to refer to the technical products which are the result of project effort.

2.6.2 Capability

Capability is the ability to perform functions or roles. It is more than the existence of the tools and products, since it includes the appropriately skilled resources to operate and manage these assets. Clearly, there is little point to having a fleet of new planes if flight crews are not trained to fly them, or the maintenance teams are not trained and certified to service and prepare them for flight, or supply chains are not embedded for the delivery of consumables and spares.

Capability relies on both the outputs being developed within the projects *and* the skilled resources to apply these outputs to the operational environment being in place. The capability requires changes to be planned or made within the operational environment to ensure that the appropriate skills, processes, and procedures are available for the transition to be effective.

The term "platform" is often used to denote the set of completed outputs ready to hand over to the stakeholders or client. "Platform" will not be used in this text, to avoid any misunderstandings between the handover of the outputs and the readiness for those to be integrated into the operational environment.

2.6.3 Outcome

Outcomes are the results of the changes which have been enacted. Perhaps the most straightforward way of considering the "outcome" is that it represents the point at which the products are being used effectively within the operational environment—it is the new operational state.

The outcome is achieved at the completion of the transition period and represents the sustainable, new business as usual (BAU).

The outcome is a desired condition and is not always defined with the same rigor as a benefit. This is a means of demonstrating that the operational environment is sustainable that is due to a number of specified users having engaged with the capability. In some cases, the outcome will represent a measurable advantage (a benefit), such as that 100 contractors are registered to use the new electronic tendering system. Alternatively, the outcome may be defined from an operational perspective: The new electronic tendering system is online, ready, and available for contractors to use.

2.6.4 Intermediate Benefits

An intermediate benefit is one which is realized as part of the overall program and is required for future benefits to be realized. Intermediate benefits are realized in a relatively short timeframe and can be measured and reported to demonstrate progress toward the longer-term results and benefits. As a result, the reporting of these benefits is likely to provide confidence to the stakeholders that the changes are effective and that the business case is likely to be realized. Alternatively, if the intermediate benefits are delayed, lower than forecast, or not realized at all, the initiative may be amended to respond to this new information.

This recognition of the importance of the intermediate benefits allows early intervention, by the sponsor and other stakeholders, to take opportunities which arise or to address underperformance. In addition, the intermediate benefits can be considered as stepping stones in the pathway to the complete set of benefits and clear indicators of the progress toward the overall targets and objectives.

2.6.5 Emergent Benefits

In a complex environment, it may not be possible to identify all of an initiative's benefits while conducting the initial planning. However, once an initiative is underway and greater understanding of the changes is gained through progressive elaboration, new benefits may be identified or observed. These are referred to as "emergent benefits."

Emergent benefits can be included in an updated business case if their impact will be legitimately included in future budgets and reporting, and they arise from the initiative. Strong governance should be applied to the claiming of emergent benefits to confirm that they have a rightful place in the business case.

2.6.6 Dis-Benefits

Dis-benefits are measurable and detrimental outcomes which result from a change initiative and are perceived as being negative or disadvantageous by one or more stakeholders. A dis-benefit is a "negative benefit." It has all of the characteristics of a benefit except that its impact is detrimental to one or more of the stakeholders. It is important to identify these dis-benefits and include them in the business case to avoid shocks and surprises later in the initiative. The inclusion of these negative consequences will, obviously, affect the viability of the initiative; however, it will set more accurate expectations and enable the sponsor to make better-informed decisions.

Examples of dis-benefits include:

- Increase in Helpdesk staffing levels for one month following the transition of all users from Microsoft® Office to Office 365. There is a definite cost associated with this increased level of service required to address the greater number of queries.
- Initial reduction in efficiency due to the need for familiarization with new processes and tools.
- Costs of additional road signage for a limited period following changes in road layouts to help drivers as they become familiar with the new road scheme.

In some cases, the dis-benefits will be addressed as costs associated within projects or the transition period. The danger with this approach is the reduced visibility of the negative impacts of change. Identifying the dis-benefits and estimating their impact will ensure that they are recorded in the business case and brought to the attention of the sponsor. This will reduce the likelihood of these consequences being covered within the operating budget of the unit affected.

Dis-benefits should be addressed in exactly the same manner as benefits, that is, the dis-benefit should have a benefit profile, and plans should be developed to manage, and minimize, their impact on the business case.

2.6.7 Benefits

Benefits are measurable improvements, in the "operational" environment which result from a change initiative and are perceived as being positive or advantageous by one or more stakeholders.

To claim that a benefit has been realized, it must fulfill five criteria:

1. Description—A clear description of the benefit must be produced.
2. Observable—A baseline must be established, which can be used as a comparison for future-state performance levels.
3. Attributable—Benefits must be generated as a result of the changes made.
4. Measurable—A method must be established for measuring the benefit.
5. Legitimate—The advantages must be legitimately realized and recognized in future budgets, metrics, and reporting.

2.7 Transition Period

The transition period follows the completion of a project, during which the "ownership" of the system is transferred from the project team to the operational team. It may be a short or lengthy period depending on the initiative. It is often characterized by a period of lower productivity as the team adjusts to the new way of working and undergoes changes to operational activities. This phenomenon should be recognized so that resources can be applied to this period of change and uncertainty, to ensure the optimization of the changes and the minimization of any negative impacts.

At the end of the transition period, the outcome will be achieved.

2.8 Transition Activities

Following the completion of the project, during the transition period, the operational team and others will undertake some activities in addition to their normal workload. These transition activities are required to ensure that the outputs delivered within the project are embedded into the operational environment and are utilized to generate the outcomes and ultimately the benefits. These activities may include:

- Training and development—Developing new skills.
- Induction—Becoming familiar with the new facilities, systems, and processes.

- Population of a new database with information.
- Cleansing and transferring data.

2.9 Baseline

A baseline is an initial measure of performance, which will be used to compare with future post-initiative performance to determine the magnitude of the benefits that have been realized. Ideally, this should be measured before the initiative begins. For programs and projects which take a considerable time to deliver, there may be merit to remeasuring the baseline closer to the transition period.

Figure 2.2 Possible Timeline for Establishing Baseline

Figure 2.2 shows a common situation with respect to the timing of projects. The concept of a project may be defined because of a need, or a perceived need, for improvement in some operational metric. Measuring a baseline at this point (Baseline 1) establishes the performance level before the project starts and before the project is widely publicized. There is often a delay between the concept being identified and resources (funding) being made available for the project to commence. If that time gap is significant, establishing a baseline at the commencement of the project may provide a more relevant reference point for comparison with the post-project performance.

Some projects merit an additional baseline measure at the end of project delivery, immediately prior to the beginning of the transition period. If the project is particularly lengthy, or if there have been changes in the operational environment during the delivery of the project, this additional baseline will record whether there has been a change in performance during the project delivery period.

2.10 Operational Environment

The operational environment, in this context, is defined as the environment or system which is changed by the initiative. It is not intended to be limited to a manufacturing or operations-based organization. In terms of an initiative, the term "operational environment" is intended to cover, but not be limited to, the following:

- Management of public spaces by a council
- Management of the flow and treatment of patients within a health facility
- Road systems and traffic flows
- Logistics and supply-chain processes
- Use of parklands and open spaces by the community
- Transactions within a financial institution
- Creation and implementation of legislation within government

The operational environment is not necessarily a corporate or government environment; it can also represent a community or not-for-profit organization.

2.11 Business as Usual

The current, approved (or recognized) natural state is referred to as "business as usual" (BAU). It represents the normal method of working or engaging with stakeholders. Within the benefits management life cycle, there will be two business-as-usual states:

- "Initial" business as usual—The way of working prior to the start of the initiative.
- "New" business as usual—The way of working after the projects and associated transition periods have been completed.

The initiative should be designed to ensure that the new business as usual is more efficient than its prior state.

2.12 Gateways

Gateways are formal review points within the initiative at which decisions are made regarding the continuing investment in the program. They should be conducted independently of the delivery team, to remove bias and to introduce a fresh perspective regarding the work which has been completed and that which is planned. As a result, the reviews will inform the sponsor(s) and key stakeholders, enabling informed decision making.

Gateways may be conducted at specific, preordained points within the benefits realization life cycle as part of a corporate governance mechanism. They can also be introduced in response to situations and circumstances which arise, so they may be imposed on an ad-hoc basis, in addition to any preplanned requirements.

Gateways may also be referred to as decision points, stage/phase gates, and milestone reviews or similar terms.

2.13 Benefits Reviews

Benefits need to be measured. This requires the cooperation of the business change manager and operational personnel in addition to those who undertake the reviews to measure the value of the benefit at that point in time. Generally, these reviews are undertaken over a relatively short period of time, during which the values of realized benefits are measured and evidence is collated to corroborate these values.

These reviews should be conducted by independent assessors and planned to optimize these resources. It is an advantage if the benefits are measured as part of the operations on an ongoing basis.

2.14 Fiscal Year (or Financial Year)

The fiscal year, or financial year, refers to the 12-month period used by the organization for accounting purposes. It coincides with the annual reporting cycle for tax purposes and the preparation of financial statements.

2.15 Front-End Loading

Front-end loading (FEL) is the practice of committing resources to the early stages of the initiative to undertake the planning and preparatory activities. "Loading" is used to infer that extra resources are made available compared to the expected norm. The loading of this phase of the initiative is designed to ensure that the workload is understood, the plans are adequate, risks are reduced, and the initiative is more likely to be successful.

2.16 Return on Investment

The term "return on investment" (ROI) is used, generically, to indicate the benefits arising to the sponsor, or other stakeholders, for a particular investment. It is not used to indicate any specific method of calculating the value of that return.

2.17 Benefit Cost Analysis

"Benefit cost analysis" is a generic term for methods of comparing the benefits generated from an initiative to the costs of delivering the initiative and the transition costs. A number of methods can be applied to undertake this analysis; some of the more commonly applied methods are discussed below.

2.17.1 Payback Period

Payback period is a simple method for measuring the overall value of an investment. The payment period is defined as the time required to recover the costs of the investment, thus reaching a breakeven or zero-sum point. The payback period is usually expressed in months or years and can be used to compare investments of similar magnitude. Generally, for comparing initiatives of similar size and importance, the "best investment" will have the shortest payback period.

2.17.2 Net Present Value

The net present value (NPV) takes into account two important aspects of the investment:

- The timing of payments (costs) and incoming revenue (benefits)—Future payments are discounted to an equivalent value of the start of the investment. This provides a fair comparison between the outgoings and the income (benefits).
- The initial investment and the ongoing costs are considered so that the benefits are an accurate reflection of the changes.

Generally, for comparing initiatives of similar importance, the "best investment" will have the largest NPV.

2.17.3 Benefit–Cost Ratio

The benefit–cost ratio (BCR) is an indicator of the overall value for money of the project, and it is calculated by dividing the total value of the benefits realized from the changes created by the project by the total cost of delivery and transition. Similar to the calculation of the NPV, the benefits value used in calculating the BCR should be expressed in discounted present values.

Generally, for comparing initiatives of similar importance, the "best investment" will have the highest BCR. A BCR greater than 1.0 indicates that the value of the benefits exceeds the costs of delivery and change. BCR is a useful method for comparing the relative merits of several investment opportunities, which can vary in size.

2.18 Diagrammatic Methods

In many cases, communication with stakeholders is enhanced through the use of some visual representations. These aids also help the team to understand the complexities of the initiative. Two of the most useful methods are defined below, and they will be discussed in detail in Chapter 6.

2.18.1 Benefit Map

One of the most effective diagrammatic methods used as a graphical aid to identify the benefits and the associated projects is benefits mapping (AXELOS, 2011). This technique applies the "start with the end in mind" philosophy and is best utilized by starting with the identification of the corporate or strategic goals followed by the final benefits which represent the metrics of those objectives. Working in reverse, from the end result toward the projects aids the identification of the appropriate projects by selecting only those which are connected to the desired benefits. The map itself resembles a network diagram of components which link projects, through outcomes and benefits, to the final desired results.

2.18.2 Dependency Network

The benefits dependency network (BDN), developed by Ward and Daniel (2006), uses five types of objects to show the pathway from initiative to long-term goals. The five types of objects are

- IS/IT enablers
- Enabling changes
- Business changes
- Business benefits
- Investment objectives

This method was developed initially for IT projects but can be applied to other initiatives. The details of this technique will be discussed more thoroughly in Chapter 6.

2.19 Documentation

A number of documents are required as part of any initiative. The major, benefits-related documents are introduced below by a brief overview. The purpose of this section is to provide an introduction to each document and provide some context for them. Detailed discussions of each document will be provided in later chapters, during the discussion of the process in which the document is first produced. The purpose and contents of each document will be presented during those chapters and summary templates are found in Appendix I.

2.19.1 Benefits Management Strategy

Each initiative may have a different approach to the management of benefits. This approach should be recorded in the *benefits management strategy*, which is a relatively high-level document which establishes the groundrules for the management of benefits. The strategy will include information such as:

- Terminology and definitions to be used within the initiative
- Specific responsibilities for members of the team
- Requirements for establishing a baseline or baselines for current performance
- Requirements for measurement of benefits
- Overall duration of benefits measurement timeframe

The strategy will not contain details of individual benefits and their metrics, because this information will be recorded within the benefit profiles and the benefits realization plan.

The strategy will contain instructions and guidelines for the delivery team and address the key questions of "what" and "why" benefits are important to this particular initiative.

Although the strategy may be tailored to the needs of a particular initiative, some organizations develop an overarching and standardized strategy, which can be applied consistently to all initiatives.

2.19.2 Benefit Profile

A *benefit profile* records all of the details of a single benefit. It is a key document because it forces the stakeholders to describe exactly what the benefit is and how it will be measured. In terms of changing the conversation about initiatives, this is the most important document, because it forces a change in the outlook about the purpose of the investment.

2.19.3 Benefits Register

The *benefits register* is a summary of the benefit profiles. It is a list of all of the benefits together with key information regarding each one. It is intended to provide an overview of the benefits expected, and as such, will not duplicate all of the details and the depth of information contained within the benefit profiles.

2.19.4 Business Case

The *business case* is a document which contains information regarding the cost and delivery timeframe of the program or project. It also contains details of the value and timing of benefits and a comparison of costs and benefits. This appraisal of the investment is the basis for justifying the initiative, and the business case document should be kept up to date as the initiative progresses, changes are made, or new information becomes available.

In addition to the document, the term "business case" is used to refer to the general justification of the investment. In this text, it should be clear from the context which meaning is intended.

2.19.5 Benefit Realization Plan

The *benefit realization plan* is more than just a schedule of points at which benefits will be realized and can be measured. That information is part of the plan, but additional detail will also be included, such as:

- Who will be involved in the identification and definition of the benefits
- Points at which the plans and business case document will be reviewed
- Details of the methods to be applied, and their timing, for reporting
- Resources required to measure the benefits
- Details of the relationships between the projects and the benefits
- Details of individual benefits (i.e., the benefit profiles)

2.19.6 Program Plan

The *program plan* is the key reference document for the initiative. It will contain the schedule, resource, and cost information. Additionally, it should contain or refer to all of the key information required for the delivery of the program, including scope, team and roles and responsibilities, controls, and plans for engagement of stakeholders. It should be updated as the work progresses. One way of considering the program plan is that it will be used to induct new team members and needs to contain all of the program information which they require.

In many organizations, the program plan is a compilation of a set of subsidiary plans. The program plan is the overarching document that holds these other documents which are linked through dependencies. The subsidiary plans include:

- Benefit realization plan
- Transition plan(s)
- Sustainment plan

2.19.7 Transition Plan

Several *transition plans* are generally required, one for each of the transitions which will take place during the initiative. Each transition plan details all of the activities, resources, and costs associated with the work required when the outputs from projects are integrated into the operational environment. It will be used, predominantly, by the business change manager to prepare for the communication of the changes and the delegation of work to all of those involved in the transition period. Some of the activities within this plan may overlap with the project, and key information will be dependencies between the project and transition.

2.19.8 Sustainment Plan

There may be activities which are required to be undertaken for the changes enacted and the benefits to continue. This work may be required outside of the initiative and could involve corporate groups or functions. A *sustainment plan* documents these requirements with timeframes and a clear discussion of the connection between the continued realization of benefits and these actions. The plan needs input from those groups who will need to implement the sustainment actions.

2.19.9 Review Report

There are a number of points at which the investment may be reviewed. With respect to benefits, these reviews will focus on the readiness for change and the value of benefits realized. The reviews, and subsequent reports, are opportunities to report successes and identify actions required to address shortfalls. Based on the information raised at these points, forecasts of future performance and the business case itself can be revised.

2.19.10 Benefits Closure Report

The *benefits management strategy* will determine the overall timeframe over which the benefits will be measured. At the end of that period there should be a final review of the total value of benefits accrued as a result of the investment. This value can then be compared with the forecasts and the business case. This is the final point at which benefits are measured as part of the initiative. The report will be the definitive and most accurate assessment of the value of benefits realized and will be used to determine the success of the investment.

2.20 Summary

The language and terminology used by the stakeholders is very important. The terms must be clear, unambiguous, and understood by all stakeholders in a consistent manner. Other terminology may be introduced to supplement, or even replace, that above. However, if that occurs, each new and additional term must be defined clearly within the team and organization.

The documentation which will be discussed as the processes move through the benefits life cycle have been introduced at this point to provide some context.

In a similar vein, the next chapter will introduce the roles within the program team and explore how each will contribute to the realization of benefits. Establishing roles and obtaining the commitment of the stakeholders to engage through their elected and appointed roles is vital to the governance of the initiative.

Exercises and Activities

1. Compare the definitions provided in this chapter with the terminology used within your organization. What other terms do you use?
2. Consider the documents listed in this chapter. Which of these are currently in use within your organization? Can some of these be combined with documents which are already used within initiatives?

Chapter 3

Team Roles and Responsibilities

> "Although some roles hold less glory, they are no less important."
> – Becky Sauerbrunn

> "All the world's a stage,
> And all the men and women merely players;
> They have their exits and their entrances;
> And one man in his time plays many parts. . . ."
> – William Shakespeare, *As You Like It*

> "Sometimes a player's greatest challenge is coming to grips with his role on the team."
> – Scottie Pippen

One of the persistent problems in managing projects is obtaining the commitment of stakeholders to undertake specific roles within the team and taking on the responsibilities of those roles. When considering programs which may require different groups of stakeholders because the costs and links to strategic goals are greater, the issues of roles and responsibilities are magnified. Programs and enterprise-wide initiatives may attract the interest of more senior and more powerful stakeholders because of the breadth of the impact of the initiative. This chapter will define the team the initiative will need, and the roles and responsibilities that must be assigned.

To be successful in implementing change within organizations or, more widely, in the community, two distinct elements have to be addressed:

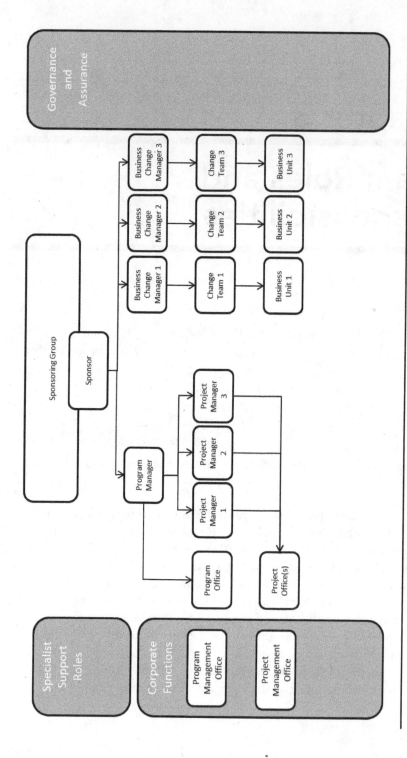

Figure 3.1 Team Structure for an Initiative

1. Delivery—The right tools need to be delivered within the time and cost constraints. This is represented by the team led by the program manager and shown on the left-hand side of Figure 3.1.
2. Transition into operations—The tools, or deliverable, must be accepted by the operational units that will deploy them. The success of the change hinges on the transition into the operational environment. The teams that will undertake these roles are shown on the right-hand side of Figure 3.1.

These two elements cannot be undertaken independently; there must be a great deal of communication between the delivery side of the team and the transition-into-operations side to ensure that the deliverables planned can be beneficially applied to the operating environment. There must be an ongoing open channel of communication between the program manager and the business change managers—if necessary, through the sponsor.

3.1 Sponsoring Group

Key stakeholders expect and deserve to have an influence in the direction of any initiatives. The sponsoring group is assembled from these stakeholders to provide that direction and leadership, which will be needed for the initiative to be successful.

The group should comprise:

- Individuals and representatives of the groups who will contribute significantly toward the resourcing and funding of the initiative
- Representatives of the groups who will be most impacted by the initiative and its changes
- Representatives of the groups who will realize the significant proportion of the benefits

As a result, the sponsoring group will contain senior members of the organizations committing to and impacted by the initiative. A key stakeholder might be a senior executive or manager with the authority to commit resources to the projects and changes included within the initiative. Some members of the sponsoring group may represent the communities affected by the delivery of the initiative or its outcomes.

The sponsoring group needs to be inclusive of the relevant parties to the initiative, which may result in a large group of opinionated and influential people being assembled. Large sponsoring groups are difficult to engage and manage. As a suggestion, the group should be limited to somewhere around seven to

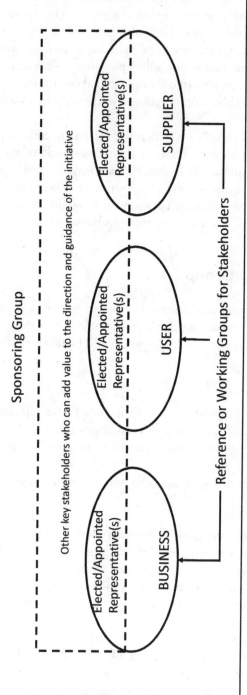

Figure 3.2 Composition of the Sponsoring Group

ten people, if it is to perform as an effective and responsive group. Reducing a large sponsoring group to a more manageable and effective size is a treacherous and delicate task, which requires strong emotional and political awareness. It may offend some influential people, but it is likely to lead to a better working environment than trying to engage over an extended period of time with a much larger group. One method which has been applied successfully to manage this issue is to group key stakeholders into the three interests identified in PRINCE2® (AXELOS, 2017), with each group appointing one or two members to the sponsoring group. Figure 3.2 highlights these three interests:

1. The business interest—Representing the perspective of those investing funding in the initiative and focused on the return on the investment.
2. The user interest—Representing the organizations, teams, and broader stakeholders who will use the capability.
3. The supplier interest—Representing the technical resources and teams who will undertake the delivery of the initiative.

The sponsoring group will be involved in the initiative when decisions regarding the investment are required and if there is a need for significant changes to the initiative. Taking into account the benefits life cycle, this will include the following occasions:

- During the context phase
- At gateway reviews
- At commissioning of each project or phase of the program
- If the business case changes, or if a decision is required regarding the business case
- By exception, if any of the tolerances set for the initiative are forecast to be breached

To make communications and instructions clear and unambiguous, the sponsoring group should appoint one member to be the primary representative for the group and its main contact. This person becomes the "sponsor." The terms of reference for this role are developed and approved by the sponsoring group. This position description of the role should define precisely the authority and autonomy of this individual.

If the right group is assembled, it will include senior, experienced, and busy professionals and stakeholders. To obtain the ongoing commitment of its participants, this role should remain a part-time function. In general, the sponsoring group should only be involved when significant investment decisions are required, which will be at the commencement of the initiative and at preplanned points to determine how the initiative proceeds, and if something significant goes wrong.

3.2 Sponsor

Generally, the sponsor will be the person who is accountable for the initiative's most significant investment; that is, the sponsor contributes the majority of the funding. He or she may also be the person who represents the recipients of the significant benefits. However, the sponsor must have the endorsement of the sponsoring group, and ultimately, it is this group which will appoint one of their own to act as the sponsor. It is possible to conduct a program without a sponsoring group, and with only a sponsor, if one person is accountable for the funding and ultimate success of the business case. Given that the nature of most change programs is transformative, and that a number of parties may contribute to the funding and be accountable for elements of the business case, it is unlikely that the key stakeholders will be adequately represented by a single person. The sponsoring group will be a strong, influential, and inclusive forum, which will direct and guide the initiative.

The sponsor will perform the following duties:

- Primary representative of the sponsoring group
- Approves the business case
- Confirms that the stakeholders will provide the resources for the achievement of the business case
- Key decision maker for issues raised by the business change managers and the program manager
- Resolves conflicts within the initiative and between the program manager and the business change managers
- Liaises, coordinates, and reports to the sponsoring group
- Monitors the business case
- Engages subject-matter experts to advise, as necessary
- Enforces governance and assurance activities

Although the sponsor needs to commit more of his or her time than the other members of the sponsoring group, this role should still be viewed as a part-time commitment to be involved in the decision-making process. It is expected that the program manager and the business change managers will address the majority of the day-to-day issues and will only escalate matters to the sponsor when they have reached, or exceeded, their designated levels of authority or if there is a major disagreement between the two roles. The sponsor will be able to plan the majority of his or her time commitments because they will coincide with predetermined approval, or decision, points and reports. They will be required to become involved, hopefully not often, by exception, but this can be optimized by the appropriate design of the other roles on the team.

3.3 Program Manager

The function of the program manager is primarily to coordinate the projects and related ongoing activities within the initiative. To achieve this, the program manager will need to develop strong working relationships with the sponsor, the business change managers, and key stakeholders within the program and its projects.

The program manager may take an active role within the project if there is a need to become involved. The program office will support the program manager by providing specific resources to undertake some of the work associated with the role. If a program management office is in place, the program manager will be able to access guidance and support from this corporate body. Ideally, the program manager should manage the interdependencies between the projects and avoid becoming technically involved in a specific project, to more effectively ensure that the program remains aligned to the organization's strategic goals and objectives and that its promised benefits are delivered.

The role of the program manager is quite different from that of a project manager. There are areas of common ground, such as the planning and coordination of the projects and their resources, and the management of risk and issues. However, the majority of the time and effort of the program manager should be spent on more strategic activities, including:

- Understanding the context of the initiative
- Liaising with the key stakeholders
- Liaising with the business change managers
- Ensuring that there is a common understanding of the timeframes for delivery and transition
- Coordinating the interdependencies between projects and the use of common resources
- Benefits realization management
- Ensuring that benefits are realized

On a day-to-day basis, the program manager is responsible for:

- Developing the overall plan for the initiative
- Developing the benefits realization plan
- Initiating/commissioning the projects

- Managing the program-level
  - Budget
  - Schedule
  - Scope and changes
  - Risks
  - Resources
  - Benefits
- Establishing an appropriate monitoring and reporting regime
- Monitoring progress of all live projects
- Addressing risks, issues, and conflicts within and between the projects
- Communicating with stakeholders and the business change managers
- Ensuring that the program will be able to deliver the capability and, ultimately, the benefits promised in the business case document
- Reporting to the sponsor on the progress of the initiative and the business case

The program manager should be a full-time role, although this will depend on the scale of the initiative and the support available. Generally, the program manager is appointed when the sponsoring group agrees to the need for the initiative and the sponsor is appointed, and the role is terminated when the final project is completed. Engaging a program manager as early as possible is a sound investment to ensure that a senior member of the team can commit time and effort to the very early planning stages, even as the business case is prepared. Merrow (2011, p. 338) declared that front-end loading (the early allocation of resources to the earliest stages of a program or project) for major projects was the world's best capital investment.

3.4 Project Manager

The program manager commissions projects in accordance with the program plan, and then each project is assigned a project manager whose role is to plan, implement the delivery of the project, and monitor the progress against the plan.

The project manager reports to and takes instructions from the program manager. Depending on the size and complexity of the project, the project manager may have support within the program manager's core team or from a project office.

The project manager is responsible for:

- Planning the project
- Assembling a team for the project
- Issuing work to the team in accordance with the plan

- Adjusting the plan based on approved changes and progress
- Managing project risks
- Reporting to the program manager
- Reporting by exception and escalating problems to the program manager
- Ensuring that the products are as specified
- Delivering the products in the required timeframe

The project manager role focuses primarily on the coordination of the technical work required to deliver the specified and agreed outputs. In any initiative there may be several projects and therefore several project managers, all reporting to the program manager.

The project manager is responsible for liaising with the users to ensure that products are defined and delivered in a way which complies with the operational environment and meets the quality and reliability requirements of the stakeholders.

3.5 Program/Project Management Office (PgMO/PMO)

"As the importance of project management has grown, many organizations have identified the need for a control center of organizational project knowledge" (Letavec, 2006). The terms *program* and *project management office* refer to those organizational bodies, or groups, which are established to ensure that there is no loss of this valuable corporate knowledge when programs and projects end, and some of the personnel leave the organization or are transferred to different roles. One of the advantages gained from establishing these groups is the development, support, and enforcement of a standard approach to the management of programs and projects.

These bodies do not exist in all organizations. Where they have been established, the program manager and sponsor has access to resources, skills, and support, which can be considered as experts and people who are familiar with the approaches taken within the performing organization. There are many different forms of program and project management offices, and the role they can play in the delivery of initiatives depends very much on the resources committed to the group: the budget to manage the group, the seniority and experience of the personnel recruited, and the authority granted to the group within the organization.

Using Letavec's (2006) model, the program management office can take three forms:

1. A *standards organization,* which will require the acquisition, or development, of
   - Templates

- Standard methodologies or techniques
- Software tools
- Reporting mechanisms
- Ranking models for project selection
- Methods for calculating benefits

2. A *knowledge organization,* "sole source of the truth" for program and project information, including
   - Key documents
   - Plans
   - Progress and schedule information
   - Benefits
   - Risks and changes
   - Lessons

3. A *consulting organization,* where the PMO will contain experienced and respected professionals who are able to advise each program on some matters, including
   - Business analysis
   - Requirements definition
   - Scheduling
   - Benefits identification, quantification, and measurement
   - Financial management
   - Risk management

Each style of PgMO or PMO requires a different set of skills but will ensure that corporate knowledge is retained and available for future investments. These roles and responsibilities can be applied equally to the program or project environment.

3.6 Project Office

The project office is established by a project manager and is designed to provide support to that particular project. The resourcing of the project office comes from the project's budget. The need for a project office will depend on a number of factors:

- Scale and complexity of the project
- Skills and experience of the project manager—which will indicate the needed amount of support
- Existence or absence of other support from the program office or corporate groups

The skills and resources of the project office will vary from project to project based on the need for specific support within the project team. However, areas commonly provided for include:

- Scheduling
- Financial management
- Risk management
- Administration

There may be advantages to pooling resources among the projects to establish a larger support group for a number of projects.

3.7 Program Office

The program office is established by the program manager and is designed to provide support to the program and the program manager in particular. Similar to the project office, the program office is established to meet the specific needs of each program, and as a result there is no standard design template. However, common roles and areas of support include:

- Benefits management
- Stakeholder engagement
- Monitoring and reporting on program-level progress
- Communications
- Configuration management and issue management
- Risk management
- Administration

The role of the program office is determined by the program manager and is funded from the program's budget.

3.8 Business Change Manager (BCM)

The business change manager (BCM) is perhaps the most influential role within the team with respect to overall success. This role is responsible for the transition from the project to operational environment and for establishing the working environment in which the benefits will realized.

There may be more than one business change manager within an initiative. In fact, there are often several active simultaneously during a complex initiative.

The role should be a part-time one, to ensure that the right person can be appointed while maintaining his or her operational role.

The business change managers must be selected from an operational role, that is, they should be the leader of the group who will be undergoing the transition and change. The appointee(s) need to have characteristics which can only come from holding a senior operational role:

- Authority—The authority to make changes in the operational environment comes from holding a leadership position in that environment.
- Credibility—It is important to engender trust in the stakeholders, that the BCM is seen as a credible, knowledgeable, and empathetic partner in the initiative.

Placing someone from a nonoperational role in this position is problematic because he or she will be perceived as an outsider—someone telling the users how to do things better. There is likely to be resistance to any changes proposed. Whereas, if the BCM is an operational lead, there will be familiarity with the users and their issues, which should make the changes less problematic and easier to promote.

The primary activities to which the BCM contributes are

- Identification of benefits
- Assessment of benefits
- Planning for transition
- Identifying and prioritization of changes
- Liaising with the program manager regarding progress of project(s) and preparations for transition
- Reporting to the sponsor
- Decision to "go live"
- Managing the transition
- Establishing a stable operational environment
- Monitoring progress toward benefit goals
- Modifying the operational environment in response to monitoring
- Maintaining the operational environment

This ensures that key stakeholders are engaged from the beginning of the initiative, reducing the likelihood of a mismatch between the products delivered by the project and those expected by the operational groups who will use them. This approach also ensures that the role remains a part-time commitment (except for managing the transition period, which is likely to be a short, intense, and full-time commitment) while remaining in contact with the key individuals in the initiative and being kept abreast of progress, proposed changes, and key date forecasts.

It is not uncommon for a BCM to be appointed for each business unit within an organization which is participating in the initiative. This is an advantage from an engagement perspective, but it may lead to difficulty in reaching common ground and agreement, particularly with the scope and extent of the initiative. There may be too many BCMs with differing priorities.

This role is difficult to fill because the skills and attributes which are required to reach the leadership role within the operational environment are quite different from those needed for the planning and management of change. However, help should be available through the change team.

3.9 Change Team

A change team is established by a business change manager and is designed to provide support to that BCM for the purpose of implementing the transition effectively. Each business change manager is responsible for designing and appointing a change team to suit the circumstances, taking into account the scale and significance of the transition and the BCM's experiences and skillset. For this reason, change teams come in all shapes and sizes. This may not be particularly helpful as guidance, but there are a number of functions which are commonly employed in support of the BCM:

- Benefits assessment and measurement
- Planning for the transition
- Communications and engagement activities
- Change management

The change teams may be appointed on a full-time basis, but more practically will be part-time commitments from each member of the team. There will be some variety in the timing of the effort, which is designed to suit the activity within the initiative and the need for involvement of the business change manager. Often, little effort will be required for some of the business units near the beginning of the initiative, and more activity will occur closer to the start of project(s).

On occasion, there may be advantages to forming a change team to support more than one business change manager, or indeed all of them. This will ensure a consistency of approach throughout the initiative and also ensure that expert resources can be recruited and devoted to the initiative for its duration.

In any case, the establishment and terms of reference for a change team is the responsibility of each business change manager.

3.10 Benefit Manager/Owner

In some instances, the BCM will be the recipient of the benefits and will have a vested interest in ensuring that they are realized. However, there are occasions when other stakeholders may be identified as having an interest in the realization of some of the benefits. A separate role may be appointed to manage the day-to-day activities involved in ensuring the realization of those specific benefits. This optional role can be used to provide support to the BCM and maintain focus on all of the benefits by assigning the monitoring of the benefits and associated activities to a benefit manager.

3.11 Assurance Roles

Assurance can be defined as: "a positive declaration intended to give confidence" (Dictionary.com, 2017). Assurance within an initiative should focus on the independent checking of information, forecasts, and monitoring. It is designed to offer an unbiased review of documentation, reports, and progress to present findings or advice to a decision maker. An unbiased confirmation that information is correct will provide the decision makers with affirmation that they can rely on that information and make informed decisions. The key attribute for someone, or a team, undertaking an assurance role is independence from the delivery team.

Within an initiative, assurance may be applied to:

- Confirming that the business case document is complete and accurate
- Checking that the progress being reported is correct
- Reviewing forecasts for planning
- Reviewing risks within the risk register
- Readiness to "go live"
- Gateway reviews

Assurance is the responsibility of the decision maker. That is, immediately before a decision is made, the decision maker must ensure that the information he or she is relying on is adequate and accurate. As a result, assurance roles can be employed on an ad-hoc basis to address the immediate needs of the decision maker. Often, the decision makers will be able to confirm themselves that the information is correct, meaning they conduct their own assurance. However, if the decision is significant, having an independent person, or people, conduct a review and confirm the results will provide direction and ammunition for the parties making the decision.

Assurance roles are considered part-time roles and are only appointed to conduct a specific function. In some instances this means that someone appointed to undertake an assurance function may be involved only on one occasion, while others may be involved several times. Appointing appropriate people to undertake this important advisory function is critical to the success of the initiative and to specific decisions and investments. The people engaged to undertake assurance roles may be selected from a number of sources:

- Business units which are not involved (or not too closely involved) with the units undertaking the change
- Corporate functions
- External consultants

Additionally, the function of assurance can be applied at all levels of the initiative to advise the sponsor, program manager, project managers, and business change managers. A risk-based approach should be applied to the provision of assurance. For example, a small decision (which will have little cost implications) may find the decision makers undertaking their own assurance by reviewing the documentation themselves to inform their decision. The review of a business case document, which will result in a decision to invest significant funds in a program, may best be undertaken by external consultants (or a corporate function, which could be the case if portfolio management is in place and there is a portfolio review board or comparable group) to avoid any internal bias and influence over the assessment to the viability of the investment. In some organizations, where portfolio management has been established, there may exist a corporate function, a portfolio review board or comparable group, which can undertake this function with an unbiased perspective.

Any assurance function confirms the accuracy of information being presented, and available, to the decision makers to enable them to be comfortable with the decisions they make. This differs from the governance function, which confirms compliance with policies and useful practices.

3.12 Governance Roles

The Governance Institute of Australia (GIA) defines corporate governance as follows: "Governance encompasses the system by which an organization is controlled and operates, and the mechanisms by which it, and its people, are held to account. Ethics, risk management, compliance and administration are all elements of governance" (GIA, 2017).

In many instances, much of the governance of an initiative is linked to corporate governance functions, which reflect the maturity and size of the organization. However, any sizeable initiative is likely to have reporting demands enforced and oversight from corporate functions that are responsible for compliance with standards, policies, and legislation.

Some of the common functions associated with governance include:

- Investment appraisal and audit
- Financial management
- Health and safety
- Compliance with corporate social responsibility policies
- Engineering or design approval—and subsequent changes
- Legal and contract management

Governance functions are often dictated, or influenced, by the industry and sector to which the performing organization belongs. There may be corporate or legislative requirements to operate under specific conditions. Governance roles conduct audits to ensure compliance with these requirements. Some sectors, such as nuclear power generation, aircraft manufacture, and financial institutions, are more heavily regulated than others—and rightly so. This results in independent tests and reviews being conducted voluntarily or at the instigation of an external body. In some instances, the sponsoring group, or sponsor, requires the involvement of governance roles, and these interventions and restrictions are detailed by this leadership group.

Governance is most likely to involve senior-level internal resources. Their involvement should be planned at specific points in the delivery of the initiative or if significant changes are required. Because of the differing needs of governance throughout the initiative, the need for skills and expertise varies significantly. Comparable to assurance, governance needs vary in terms of skills, expertise, and number of people involved, and are usually a part-time function. However, governance needs and demands must be included in the strategies which define the initiative, making the associated activities easier to plan.

3.13 Specialist Support Roles

Even with a program management office, and program office, there may be a need for input from a specialist with expertise to inform or enable decision making. As a result, the following optional roles may be considered.

- ***Risk manager.*** A risk manager may be engaged to provide an initiative-wide, or portfolio-wide, perspective of risks, with the ability to collate and

review the overall impact of the complete set of live risks. This will expose common risks which may be resolved at a system or organizational level. It will also highlight the dependencies between risks and the activities within the initiative.
- *Change control board (CCB).* This group may be established to consider all requests for changes within an organization or an initiative. The advantage of creating a single group for this important task is that it raises the visibility of requests for change to a higher level, ensuring that the impact of change is understood across all projects and initiatives. Change controls boards can act in two ways:
  1. Advisory—The CCB may review all requests for change and advise the sponsor on the validity and priority of each request. The sponsor then makes the final decision whether to approve the request.
  2. Decision making—The CCB may be allocated a budget which they can apply to requests for changes which it approves. In this situation, the CCB reviews all requests for change and approves them according to their terms of reference.
- *Procurement.* The organizational processes for creating and issuing tenders and managing the selection of the appropriate suppliers is often complex. Program and project managers often benefit from support in this area, to ensure that the award of contracts and acquisition of resources, services, and materials are conducted ethically and are in compliance with legal requirements.
- *Scheduling and monitoring.* Complex initiatives may require a master schedule to be prepared and maintained. Additionally, if specific tools and techniques are applied to the planning and monitoring of the progress of the work, a single, centralized point of contact will be advantageous for continuity and consistency. This may be the case if earned value management, for example, is the preferred method of performance measurement.
- *Business analysis.* This role is often found in the program office, The advantage of including business analysis (BA) as a centralized resource is that the role can be deployed to all projects within the initiative to ensure a consistent approach to identifying benefits and opportunities for change. The operational teams then have a single point of contact to discuss their changes, and the BA role develops growing domain knowledge of the operations.
- *Benefits manager/coordinator.* Benefits realization management is a new and developing area, and there may be insufficient understanding among the organizational teams to implement a disciplined approach to change and the realization of benefits. A centralized resource can provide dedicated time and support to the BCMs and change teams, ensuring that a common approach is undertaken across the organization.

3.14 Additional Considerations

Tailoring the team to suit the circumstances of a particular initiative is always a consideration. It may be an expensive luxury to have all of the identified roles allocated and resourced for the duration of the program, particularly for smaller organizations and initiatives.

A risk-based approach should be adopted when considering any tailoring, especially if it deviates from good practices. If this approach is implemented, it results in resources being allocated to areas which have a greater risk of adversely affecting the impact of the investment. Lower-risk areas of the initiative may have fewer resources deployed, with the expectation that one member of the team will undertake some activity on a part-time basis to cover that area.

For example, assume a smaller initiative cannot engage a dedicated risk manager, and the program manager has a background in risk management. The program manager could take on the role of the risk manager. For a larger initiative, this additional workload may overwhelm the program manager, and engaging someone to take on this role may reap rewards because he or she will have the time and expertise to devote to the role without interference from other responsibilities.

In assembling the team, consider the benefits of implementing different structures. Use a program structure only if there are additional benefits to be gained by using that structure compared to managing several independent projects. Similarly, consolidating some roles may be necessary, due to resourcing and funding constraints. The temptation to merge roles should be tempered by the need for effective assurance and governance functions and take into account the fact that overzealous consolidation may increase risks.

3.15 Summary

The team structure and the allocation of credible and authoritative people into the roles are crucial to the smooth progress of any initiative. In many cases, the roles people take on during one initiative may be different from the roles they played in the past. Stakeholders and team members should be appointed to their roles based on their ability to contribute to the initiative. This means that team members may not necessarily be appointed to a role within the team based solely on their position within the organization. However, in many situations, selection of the team may be compromised because of the need to involve those with the authority to allocate funds to the initiative and to change the operational environment.

The next chapter will introduce the benefits life cycle, which will be detailed over the succeeding six chapters. This part of the book will explain the processes, activities, and tasks which will be undertaken by the team members as the initiative progresses.

Exercises and Activities

1. Consider the teams with which you are familiar. How were these roles applied? Analyze how the roles could have been applied more effectively.
2. How could the roles described here be integrated into your organization's change teams?
3. Discuss the use of written role descriptions and how they might be used to establish teams.

Part Two

The Benefits Life Cycle

Chapter 4

Introduction to the Benefits Life Cycle

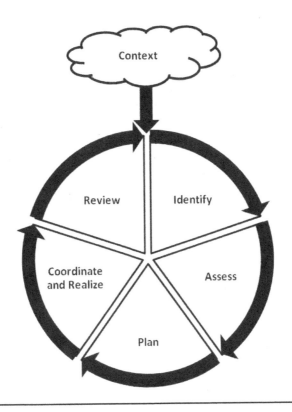

Figure 4.1 Overview of the Benefits Life Cycle

Benefits must be considered at the beginning of any investment. The benefits must become the focal point for the stakeholders, and the team as a whole, to ensure that decisions are based on the needs of those stakeholders and the eventual success criteria for the investment. The benefits will not materialize from the ether; they must be planned and nurtured and reviewed throughout the life of the initiative. To guide the actors in the initiative, a *benefits life cycle* has been developed, and shown in Figure 4.1, to reinforce the importance of action and repeated action.

The benefits life cycle includes six steps:

1. **Establish the context.** The process aims to determine the organizational and initiative environments within which the change will occur. A greater understanding of the context will assist the team's focus and delivery.
2. **Identify the benefits.** The individual benefits are identified and defined in clear and unambiguous terms. This is a valuable step in engaging the stakeholders by ensuring a common understanding of the investment's goals and purpose.
3. **Assess the benefits.** This process determines the scale and significance of the benefits as a whole, and results in the compilation of a business case document.
4. **Plan for benefits realization.** The detailed links and dependencies between projects and transitional activities and ultimately the benefits are planned and the resource requirements for transitional and other activities estimated and scheduled.
5. **Coordinate and realize the benefits.** This process covers the coordination of the tasks to be undertaken following the completion of the projects, including the transitional and subsequent activities necessary to integrate the products into the operational environment.
6. **Review the initiative.** This process is designed to review the progress made toward the initiative's goals by reviewing the value of the realized benefits. This will form the basis for required, or suggested, changes in the benefits realization plan, transition plans, and sustainment plans to ensure that future benefits are realized and sustainable.

Benefits realization management needs to be addressed throughout the initiative, and this life cycle should *not* be considered as a lineal pathway, or sequence of steps, which can be followed once in order to reach success. Rather, it should be viewed as an interdependent flow where information and findings in one step will inform others and determine which of the other steps should be applied next. For example, during step 5, coordinate and realize the benefits, a new emergent benefit may be identified, at which point some actions from step 2, identify the (emergent) benefits, should be undertaken to fully define this new benefit, before a plan can be developed for its realization and assessment.

The benefits life cycle should be considered a dynamic environment which responds to circumstances and information as the work of the initiative progresses. This is particularly true of long-term initiatives, which may be subject to internal and external influences as time goes by. It is hoped that the context is the one constant in the life cycle. If the context changes, this will have major ramifications on the initiative and the management of the benefits. However, even with the best of intentions and planning, long-term initiatives may be affected by other factors, such as:

- Changes in market or sector conditions
- Changes in personnel among the key stakeholders
- Legislative or regulatory changes
- Mergers and acquisitions

The following chapters will discuss each of the steps in the benefits life cycle in more depth, along with practical guidance regarding the implementation of each step.

Chapter 5

Establish the Context

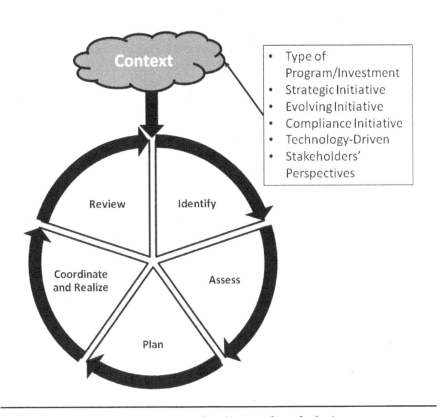

Figure 5.1 Establish the Context within the Benefits Life Cycle

> *"In this, as in all things, context is key."*
> – "Matt Santos," on *The West Wing*

> *"Context is all."*
> – Margaret Atwood, *The Handmaid's Tale*

> *"Wisdom is intelligence in context."*
> – Raheel Farooq

The context of any undertaking determines the environment in which it is delivered, which is the reason for Figure 5.1 highlighting "Establish the Context" as the starting point for the benefits life cycle and set apart from the other processes. The context of the initiative defines the background and the thinking behind the decisions to start and continue with proposed changes. The context determines the objectives and priorities which are set, and the manner in which decisions are made. It is therefore important that the program manager and the key stakeholders have a common understanding of that context, the drivers to commence the work, and how this environment influences the decision making within the initiative and particularly those decisions which affect the achievement of the benefits. This chapter will discuss the importance of establishing a common understanding of the driving forces behind the initiative.

For example, if new legislation has been enacted which affects an organization, the primary focus of any project is likely to be compliance with this legislation. This is the context and establishes the priority for any objectives within the project. It could be argued that compliance will result in the avoidance of fines or penalties associated with the legislation, and this is a measurable, and financial, benefit. However, the primary purpose of undertaking the change remains the need, or desire, to comply with the new laws and regulations. In a case such as this one, the organization and key sponsors may be more interested in the outcome (i.e., the new method of operations complying with the legislation) than in the potential for future penalty avoidance.

5.1 Drivers for Programs and Investments

The type of investment, or program, being undertaken has a significant influence on the context and the expectations of the stakeholders. The type of program influences the perspective of the team and stakeholders regarding the benefits and outcomes expected and their relative importance.

5.1.1 PESTLE

An analysis of the Political, Economic, Social, Technological, Legal, and Environmental (PESTLE) factors can assist in identifying the drivers for the investment, which, in turn, will indicate the stakeholders' primary areas of interest. These drivers will identify the likely sources of benefits, or at least those benefits which are of real interest to the key stakeholders.

PESTLE categorizes the drivers for the funding of change and programs as:

- **Political**—Internal, local, national, and regional politics may provide the driving force required to initiate an investment. Government-backed incentives make some investments more attractive, or even viable, and attract stakeholders to projects.
- **Economic**—the cost of borrowing or exchange rates may influence the viability of some projects. High unemployment may make some key skills available, or less expensive, which could lead to increased investment within an organization or region. There may be opportunities to reduce costs of operations while maintaining the standard of service provided. A project which has an economic driver is likely to focus primarily on financial issues and will seek a financial return.
- **Social**—Aging populations may be a concern to a government, making one of the desired benefits from a program the increased employment of people over the age of 55. Social programs expect some economic return, but the focus is on social issues, for example, reducing the incidence of unacceptable behaviors, such as young drivers speeding. The force behind the initiative is a social issue, which the initiative will address through changing the behavior of drivers.
- **Technological**—The availability of new technology at reasonable cost may enable projects which were not previously viable. For example, new drilling technology may open up opportunities to exploit new oil and gas fields, which were previously too expensive to explore and exploit. The opportunities may also involve risks, and these must be considered when examining the viability of the investment. Early adopters of technology may have to contend with high operating costs or poor reliability, which would be considered part of the cost for exploiting the opportunity.
- **Legal**—Investments driven by legislation are often compliance-related or arise from the opportunities created by the introduction of new legislation, such as changing planning regulations. Legislation or elections may also give rise to opportunities for an organization. For example, changes to the legislation regarding waste management and disposal may lead to increased fees for landfill. A reduction in the volume of waste produced may be an immediate benefit, which will also lead to reduced operational costs (or at least avoidance of increased costs), and this might lead to an initiative to reduce the volume of waste produced. Although the benefits may be measured in financial terms, the driver for this change is fundamentally driven by the legal changes.

- **Environmental**—An organization may decide to change its processes and material resources to ones which are less damaging to the environment. A chemical needed in the production process may be replaced by a biodegradable one, which will have a smaller, and shorter-lived, impact on the environment. The biodegradable option may be more expensive, but making the choice to use it will demonstrate that the producer is serious about managing its impact on the environment. The outcome will be raising the "green credentials" of the organization, which could attract a new and increased demographic to the customer profile (which will yield longer-term benefits).

Identifying the drivers for the investment allows the team to identify the key stakeholders, understand the motives for the change, and prioritize the benefits which are identified. This contextual information will allow the program manager to appreciate the position of the key stakeholders. The drivers will indicate what benefits may be prioritized and valued by the stakeholders. In addition, knowledge of the reasons behind commissioning the initiative and the need for change will focus the stakeholders and avoid distractions which might cause "scope creep" and the allocation of resources to actions which are not part of the initial purpose of the investment.

5.1.2 Triple Bottom Line

Many organizations in the public and private sector use the term *Triple Bottom Line* (TBL). This indicates that the organization considers three factors when assessing the success or failure of a program; generally, they are listed as:

- Economic—Representing the financial aspects of reporting, including the investment value and its comparison with financial returns.
- Social—Covering social aspects of the investment which will be valued within the community, such as greater diversity, a safer community, or a more liveable city.
- Environmental—Addressing benefits which are related to contributions to environmental impact, such as pollutant emissions, noise reduction, and reduced salinity of river water.

The organization will judge investment success based on the value of each consideration; often these will not be considered as having equal weight. This will enable a business case to be developed which takes into account all of these factors and places the appropriate emphasis on each when making decisions regarding the initial investment, and progress once the initiative has started. Each organization, and each program or project, may assign different relative importance to each of these three elements. Having a greater understanding of these priorities and the context of the Triple Bottom Line will enable the team to design the initiative and deliver it to realize the desired result.

For example, a city council may have a mission to reduce the cost of services and be environmentally friendly. As a result, the council decides to install solar panels on all of its buildings with a roof area above a minimum threshold level. The installation and commissioning of the solar panel system will result in:

- Reductions in operating costs at those locations, through the use of solar power
- Increased revenue to the council, because the excess power generated may be sold to the power company if it is not stored in batteries
- Recognition that the council and the community can be seen to be progressive and "clean"
- Reduction in carbon emissions as a result of council activities

In this case, the council needs to decide which of these results are relevant and deserve to be measured and publicized. It may be that, depending on the cost of installing the system and the price of electricity, the project does not provide the return on investment normally expected from a capital investment. However, the financial argument is only one part of the business case, and other aspects of the change may be considered in making the decision to invest in the new technology and system. This example exhibits all three aspects of the Triple Bottom Line:

- Economic—Operating costs will be reduced due to decreased electrical power needs from the grid. Revenue may even increase if an excess of solar power is generated and this power is sold to the power retailer.
- Social—Funds previously allocated to power costs can be diverted to other projects or services, such as parks, dog exercise areas, or increased library services.
- Environment—Lower power requirements will lead to reduced carbon emissions.

The financial information which will form a major part of the business case for this scenario is shown in Table 5.1.

In addition, there are benefits associated with the generation of "green" energy compared to the previous regime; it is a straightforward task to calculate the reduction in greenhouse gas emissions as a result of the implementation of

Table 5.1 Financial Information for the Solar Initiative

Cost	Benefit
Installation and commissioning costs	Reduced electricity bills
Maintenance of the solar panel system over its lifetime	Additional revenue from sale of excess power generated

the solar power system. This reduction in emissions is easy to calculate, and, as such, can be claimed as a benefit. Financial incentives may be available for reducing the production and emission of damaging gases, which would enable the council to convert the resulting reduction of emissions into a financial benefit. While this conversion will simplify the business case, it is not strictly necessary—it is acceptable for the business case to provide an explanation of the contribution of a nonfinancial element.

The value of considering a mix of financial and nonfinancial benefits is that it demonstrates the breadth of its impact, and it legitimately strengthens the business case for the investment.

In some instances, converting a social or environmental benefit into a financial item may confuse the primary issues and be distasteful to some of the stakeholders. For example, when making safety improvements to a stretch of road which is a known accident hotspot, the primary objective is to reduce the number of accidents, injuries, and deaths on that stretch. Although there are financial implications associated with traffic accidents, most people (and in this case the community are key stakeholders) would baulk at the thought of placing a value on the cost of life and injury, and it would be clearer to measure the reduction in incidents as the benefit.

5.2 Other Program Types

Managing Successful Programmes (MSP®) (AXELOS, 2011) identifies three types of programs:

- *Vision-led,* whose characteristics are that the program was commissioned to deliver on a specific and clearly defined goal, which is generated out of the corporate strategies and objectives. These programs generally are launched from the executive levels of an organization and as such tend to have strong support from senior stakeholders.
- *Emergent,* which tend to arise from concurrent projects and programs being melded together. Words which are generally used to describe their conception include "evolved" and "morphed." Although these are common, as organizations rationalize their capability and mature with respect to

project, program, and portfolio management, they tend to bring the baggage from their previous lives, and there are often issues bringing the teams together to work collaboratively on something, which may be ill-defined.
- *Compliance,* which are often considered "must do" initiatives. For some reason, often related to legislation or regulations, the organization needs to change its operating environment. There is often a legal driver to the investment, which often places the focus of the work on conformity or compliance issues. Many initiatives fall into this category. Unfortunately, in practice, this manifests itself as a concentration of effort on the outcome (a compliant environment), with too little thought given to the opportunity to generate benefits from the investment. A financial institution may be required to keep records and report to the taxation agency in a particular manner. In developing the systems to comply, the institution could take the opportunity to improve the information available to its customers, providing added value to them.

For the sake of greater inclusion of more investment types, the terminology used throughout this book will be modified somewhat to:

- Strategic initiatives
- Evolving initiatives
- Compliance initiatives
- Technically driven initiatives

The different categories of programs do not necessarily mean that a different approach for managing and controlling the program is required. There may be different priorities for each category, or different emphasis on the governance requirements and benefits focus. Recognizing that there are different categories, each with differing priorities, will help to understand the context and the stakeholders' perspectives and their attitude toward benefits realization.

5.2.1 Strategic Initiatives

Strategic initiatives are driven from senior levels within the organization and have the following characteristics:

- Led by a vision or long-term objective
- Contributing to the long-term aims of the organization
- Sponsored by a senior executive and strongly supported by the highest levels within the organization
- Often innovative

These are often exciting and enterprising programs and projects, which can energize the team and stakeholders. Generally, the objectives and focus of them are clear and unambiguous, which provides the team and its stakeholders with common goals. Because of this emphasis, these initiatives are often easier to deliver, based on the clarity of the goal, high-level support, and purpose. These facets are sometimes lacking in the program and project environment, and they have been identified as factors which are essential to success.

Examples of this type of initiative include the establishment of a new enterprise or business, such as Virgin Galactic or Hyperloop. Virgin Galactic is a relatively new organization devoted to space tourism. In such a case, a new organization has to be developed with new technologies and processes. There are few barriers resulting from existing corporate structure and legacy issues, and there is a desire within the team to create something of value. A Hyperloop has been proposed to connect Los Angeles and San Francisco with a passenger and freight train system, operating within a vacuum tube; capable of reaching speeds of 750 mph (1,200 km/h), it will cover the 400-mi (500-km) distance in about 35 minutes.

5.2.2 Evolving Initiatives

Evolving initiatives are often described using terms such as "emerging" or "morphing," for example, "The program morphed out of a number of existing projects." The programs and projects are created by combining a number of existing initiatives and are likely to have a loosely defined brief in an attempt to accommodate all of these existing projects. There should be a strong and attractive advantage to merging the initiatives, in terms of coordination or delivery costs, for centralizing the management of the projects. It is to be hoped that there would be a common, or similar, purpose behind the program, which will make the alignment of the outcomes, objectives, and benefits more straightforward and clear to the participants. Problems can occur when the purpose of establishing one evolving initiative from those already underway is primarily convenience and administrative simplicity. There is often disruption to the teams and the progress of the initiatives and objectives when merging occurs.

Evolving initiatives may exhibit some of the following attributes:

- A number of projects may be bundled together because they have a common theme.
- Several diverse projects may be merged, creating a broad, and not always connected, set of objectives.
- Large and diverse stakeholder groups with different interests may be involved and require engagement.

- Vague objectives and ill-defined benefits may result in work being undertaken which does not contribute to the organization's long-term goals and objectives.
- Uncertainty and hostility among the vested stakeholders regarding the priorities of the initiative may occur, requiring more time by the sponsor, program, and project managers to overcome or at least neutralize.

One of the major drawbacks of this type of initiative is that two or more existing teams are already committed to a goal and have developed their own team identity and culture, which may be disrupted through this evolution. This will almost certainly require a reassessment of the goals and targets of each team to determine the collective priorities. There may be a resultant lack of cohesion and synergy between the participants, not just the delivery teams, but also among the stakeholders.

However, from experience and in practice, this is a common issue. Consolidation of efforts across an organization is one of the indicators that the organization is adopting a portfolio- or program-based approach. As such, evolving initiatives need to be recognized as a legitimate approach, and the associated issues must be recognized and addressed. Some of the issues are

- Different cultures within the teams being incompatible
- Stakeholders' expectations not being met by the consolidation
- Divergence of benefits rather than focus
- Lack of a baseline for comparing benefits
- Potential for a significant number of emergent benefits

Some of these issues will have positive outcomes and may be resolved with relative ease—but they do need to be recognized and addressed proactively. Essential to the achievement of the realization of benefits is a detailed understanding of the environment and establishment of a set of agreed-on benefits for the new (or developing) initiative—in other words, a common understanding and agreement of the context and environment for the investment among the stakeholders.

5.2.3 Compliance Initiatives

In an increasingly regulated and controlled world, organizations must comply with much legislation. As these laws and regulations are revised and updated, organizations will commission initiatives to retain currency with the appropriate rules and guidelines. The fundamental purpose of a compliance initiative is to make the necessary changes to maintain compliance. There is a greater focus on the outcomes (compliance) than there is on the realization of benefits.

Compliance projects demonstrate the following characteristics:

- Forced on the organization and team, and as a result it is often difficult to motivate the team—certainly not in the same way that motivation is possible with a strategic program
- Singular focus on meeting the requirements of the new regulations or legislation
- Focus on the compliance outcomes, not the longer-term benefits
- An attitude that the most important thing is to do enough, and only just enough, to "get across the line," or comply. There is often not an opportunity to examine improvement opportunities, making the identification and realization of benefits difficult.

It can be argued that the benefits are the revenues from being able to continue to operate and the absence of fines levied. There is some degree of truth in that argument. However, this is not necessarily an engaging or exciting selling point for the teams who will be affected by the changes to the operating environment.

To engage the stakeholders and team, it is often valuable to connect the outcomes with benefits associated with larger and longer-term strategic goals. Showing that these outcomes are closely linked and even essential to the achievement of other benefits is one way of proving the value of compliance. These outcomes are often referred to as "enablers." This is a method of justifying the lack of interest in disciplined benefits management by arguing that the outcomes are a necessity, which cannot always be measured. The argument continues that eventually the compliance outcomes will enable benefits to be realized through the continued operations and other projects which follow.

The rarely asked question with respect to compliance is "How much is the sponsor willing to spend to achieve it?" The investment is being made to achieve compliance, an ill-defined target. A budget will be set aside for the initiative, but how much extra funding will become available if the goal is not met or appears to be more difficult to achieve than first thought? There is a danger that the costs, and authorized budget, will expand to allow the required outcome to be reached. A key question is "At what point will the sponsor halt funding the work?" These are difficult concepts because there is often the view that compliance is a "must have" condition and a binary one (either the team complies or it does not). There may be degrees of compliance, which may influence the investment and business case. Within the environment of a compliance project, there may be benefits which can be realized as part of that objective. For example, while upgrading the financial management system to comply with new and complex taxation legislation, changes could be introduced which reduce the time committed to other transactions, thus saving time throughout the year.

5.2.4 Technology-Driven Initiatives

Programs and projects which are driven by technology are common. They are specified in terms of the deliverables which will be produced, and there is often a gap in understanding of the objectives between the team who are delivering the work and the "client" and their team, who will accept the products and use them in operations. This environment is characterized by:

- Focus on the technical delivery of the products
- Lack of operational input to decision making
- Lack of representation of the operational personnel within the team and stakeholder groups
- Changes in the operational environment not resulting in changes within the delivery of the initiative
- The technical design and subsequent changes to the design not taking into account the operating environment
- The team's belief that all problems must be resolved by a technical solution
- The program and project are dealt with as an isolated system

There are a number of dangers with this approach because of the isolation of the delivery team. There is often a lack of interest in the operational environment while the team focuses on the technical work. The need to change the operational environment is overlooked, with no one taking responsibility for the necessary transitional activities. As a result, it is likely that the forecasts made for benefits will be missed, and opportunities will not be taken. To counterbalance this effect, the delivery team should engage with the operational groups to ensure that there is harmony between their respective agendas. There may be an opportunity to include work and outputs which may be of significant value to the business units. Creating some small, value-adding features to the technical system may offer benefits within the operations at little cost.

5.3 Recognizing the Stakeholders' Perspectives

The project team members are charged with developing an asset or capability which will affect the stakeholders and investors, and the manner in which they conduct their business or operations. The project team cannot focus only on the internal project environment; they must also understand the stakeholders' view of the project. Benefits realization management cannot be separated from stakeholder engagement. In fact, the management of benefits is a valuable tool in effective engagement of the stakeholders.

The program manager, and the other team members, should spend time during the earliest phases of the program and associated projects to gain a greater understanding of the objectives from the stakeholders' point of view. Gaining a more detailed understanding of the context of the project and its drivers will enable more effective communication with the stakeholders.

The engagement of stakeholders will be revisited throughout the following processes, to provide some tools and techniques which will be required to communicate with stakeholders and create a working relationship and will support the management of benefits.

5.3.1 Getting to Know the Stakeholders

There are few options open to the initiative team, other than to spend time with the stakeholders so as to understand their perceptions of the purpose of the initiative and its impact on them. This can be achieved through a number of means, including:

- Formal workshops
- One-on-one discussions
- Interactive meetings
- Shadowing key stakeholders

5.4 Documentation

The initial conversations between the program manager and the sponsoring group will establish the foundations for the initiative, which will be recorded in the benefits management strategy. This will provide the background and intent of the program and how it is to be managed.

5.4.1 Benefits Management Strategy

Each initiative may have a different approach to the management of benefits. This approach should be recorded in the benefits management strategy, which is a relatively high-level document. It provides guidance and instruction from the sponsoring group to the program manager regarding the conduct of the initiative and establishing the ground rules for the management of benefits. The strategy will include information such as:

- Terminology and definitions to be used within the initiative—which may be specific to the organizations involved or to the initiative itself.
- Specific responsibilities for members of the team—including accountabilities and responsibilities for the business case.
- Requirements for establishing a baseline or baselines for current performance—at what point(s) in the delivery of the initiative will baselines be measured, and, if necessary, revised.
- General reporting requirements from the initiative to the sponsoring group.
- Details of mechanisms for escalating issues and risks to the sponsoring group.
- Requirements for measurement of benefits—indicating when reviews will be conducted.
- Overall duration of benefits measurement timeframe—stating the point at which the final benefits review will be conducted.

The strategy will not contain details of individual benefits and their metrics, but this information will be recorded within the benefit profiles and the benefits realization plan.

The strategy contains instructions and guidelines for the delivery team and addresses the key questions of "what" and "why" benefits are important to this particular initiative. Although the strategy may be tailored to the needs of a particular initiative, some organizations develop an overarching and standardized strategy which can be applied consistently to all initiatives.

Table 5.2 identifies the key activities which must be undertaken as part of the process. It should be noted that the program management approach may cover several of these activities, such as:

- Assembly of the sponsoring group
- Identification and appointment of the sponsor
- Appointment of the program manager

5.5 Summary

Benefits management is heavily influenced by the environment within which the project is delivered. Having an understanding of the context of the project and the view the stakeholders have of the program or project and, more important, the benefits which will be realized following its completion, will enable the discussion between the project manager and the key stakeholders. With this knowledge, the project manager will be able to advise the stakeholders of

Table 5.2 Activities Conducted within the "Establish the Context" Process

Activity	Purpose	Responsibility	Documentation
Assembly of the sponsoring group	Identify the key stakeholders	Sponsor and sponsoring group	
Identification and appointment of the sponsor	Appoint the key decision maker and representation of the sponsoring group	Sponsoring group	Terms of reference for the sponsor (optional)
Appointment of the program manager	Appoint the key, full-time manager to plan and coordinate the initiative	Sponsor	Terms of reference for the program manager (optional)
Workshops between the program manager and the sponsoring group	Understand the context and environment within which the initiative is taking place	Program manager	
Develop the benefits management strategy	Document the key stakeholders' understanding of the objectives	Program manager, approved by the sponsoring group	Benefits management strategy

how changes to the project will impact the overall performance criteria. The categorization of projects and programs is an important technique in which assumptions can be confirmed or corrected, and the events or factors which instigated the projects identified. These will provide indicators to the program manager of the relative importance of risks and issues within the project and allow the execution, management, and governance of the project to be addressed in an appropriate way.

Some of the issues that should be addressed at this stage include:

- Is there clarity among the stakeholders regarding the general purpose of the investment?
- Have the drivers of the need for some change been identified?
- Is there alignment between the drivers and the broader organization objectives?
- Do the project manager and project team understand the environment well enough to advise the stakeholders?
- Do the stakeholders agree on the priorities for focus?

Exercises and Activities

Consider an initiative within your organization or one which is familiar.

1. Identify the drivers for the initiative using the PESTLE model. Why is the initiative important to the organization?
   a. Political
   b. Economic
   c. Social
   d. Technological
   e. Legal
   f. Environmental
2. Find examples of different types of programs and explain the characteristics which are important in their categorization:
   a. Strategic initiatives
   b. Evolving initiatives
   c. Compliance initiatives
   d. Technology-driven initiatives
3. Investigate the approach to the early engagement of stakeholders within initiatives. Present a critical analysis of this approach with recommendations for improvement.

Chapter 6
Identify the Benefits

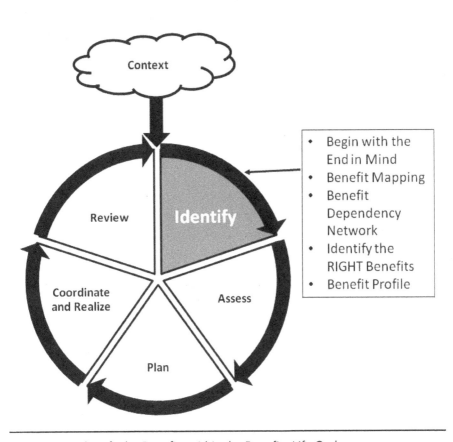

Figure 6.1   Identify the Benefits within the Benefits Life Cycle

> *"If you don't know where you are going, any road will get you there."*
> – Lewis Carroll
>
> *"Be sure you positively identify your target before you pull the trigger."*
> – Tom Flynn
>
> *"Define success on your own terms, achieve it by your own rules, and build a life you're proud to live."*
> – Anne Sweeney

Great care must be taken when identifying the benefits desired from any investment, because they will shape the nature of the initiative and the behaviors of the delivery teams and those people working in the operational environment, which will result in the changes required to realize the benefits. The benefits identified and publicized as the desired result of initiatives will shape stakeholder expectations and their criteria for success, as indicated in Figure 6.1.

This chapter will cover techniques relating to the identification of benefits and the linking of strategic objectives to final benefits and projects.

6.1 Getting Off to a Bad Start . . .

From experience, many examples could be cited where the first question asked by stakeholders is "What projects are we going to do?" This is especially prevalent in emergent programs, where work might already be underway in some projects. While this action-oriented question may seem reasonable, a better question would be "What are the benefits we want to realize?"

Stephen R. Covey, in his book, *The 7 Habits of Highly Effective People*, identified the Habit of Vision, which dictates that effective people (and teams) "Begin with the end in mind" (Covey, 1989). In practical terms this means that the context should be established first, with the longer-term objectives of any endeavor understood before the work commences.

Too often the focus is on the immediate workload—"What do we have to do?"—rather than "What do we want to achieve?" The longer-term goals must be brought to the forefront of the stakeholders' minds to set the expectations from the start of the initiative. Once an exciting idea has been put forward, it is tempting to rush head-long into activity—to start getting something done. This is the trap of *busyness* over *productivity*. Much effort can be diverted into being busy while leading the team in the wrong direction. It is comforting to see the team being active—supervisors and managers like to see activity.

To achieve the goals of change (or any business case), the teams must be successful in two ways:

1. Delivery of the projects and the change must be efficient: The teams must deliver in a productive manner.
2. Benefits must be realized, meeting or exceeding the forecasts.

Failure to achieve either of these will undermine the business case. If the delivery of the project and change is inefficiently completed, the business case will rely on increased benefits to balance the books. Similarly, if the benefits are not realized in their entirety or are below the estimated values, the business case will not be realized. Unfortunately, if the problem is the benefits being less than forecast, there is little that may be done to rectify the situation. It is too late at that point to take corrective action.

So, for success, the team and the stakeholders must be well prepared. This requires work to be undertaken at the beginning of the project to define it, its objectives, and the plan for delivery and achievement of the goals of the change.

6.2 Begin with the End in Mind

A number of visual/diagrammatic approaches can support the identification of benefits. These tools also demonstrate how the benefits support the longer-term objectives, and how each of the projects contributes toward each of the benefits.

The diagrams produced by these techniques are useful in showing the dependencies between the projects and between the projects and each of the outcomes and benefits. Simple graphical representations are helpful ways of presenting information to all stakeholders for three reasons:

1. Benefits which are of interest to the stakeholders are shown clearly.
2. There is little technical information cluttering the story behind the decisions made in selecting the projects.
3. A clear path shown connecting the projects, through outcomes, to the benefits. It is easy to see how taking action within one project will lead to one or more benefits.

When applying any of these techniques (which will be detailed shortly), it is important to identify longer-term strategies and objectives, which will be addressed first—that is, the context. The teams next identify the final benefits and then the intermediate benefits and outcomes. This process of starting with the end in mind will ultimately assist in identifying the appropriate projects, which will contribute to the outcomes and benefits desired.

6.3 Diagrammatic Techniques

6.3.1 Benefit Mapping

One of the most effective diagrammatic methods used as a graphical aid and as a method to identify the benefits and the associated projects is benefits mapping (AXELOS, 2011). This technique applies the "Start with the end in mind" philosophy and is best utilized by starting at the right-hand end of the map and working backward. This will ensure that there is a clear link between the benefits and the projects selected to realize them. Working in reverse, from the end result toward the projects, is not a familiar approach, but it aids the identification of the appropriate projects by selecting only those which are connected to the desired benefits. The benefits map in Figure 6.2 identifies each of the elements in the chain between the project and the overall corporate objectives.

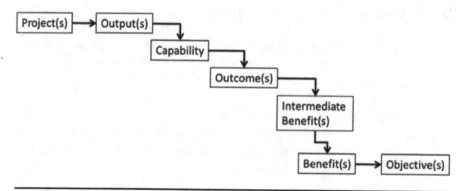

Figure 6.2 The Journey from Project to Benefits and Objectives

When applying this technique, it is best to start by identifying the objectives, which are the focus of the change, and then work backward to identify the relevant projects:

- Corporate or strategic objectives—The sponsor of the initiative and the key stakeholders will be able to identify these objectives. Hopefully, this group will be in agreement regarding the focus of the investment. These objectives may not be documented or comprehended in terms which are measurable. For example, the objective of a city council may be to create a "vibrant city." Unfortunately, this term, "vibrant city," may be interpreted in many different ways. So, the next crucial step is to decompose the objectives into measurable terms—"benefits."

- Benefits—The measurable advantages which the key stakeholders can relate as being representative of the objectives. The final benefits will be measured and clearly indicate that the objectives have been achieved. In the case of the "vibrant city" objective, benefits could include:
  - Increased pedestrian traffic in the city
  - Increased number of events held in the city
  - Reduced vacancy rates in retail premises
  - Increased number of tourists
- Intermediate benefits—These benefits are stepping stones toward the longer-term goals and do not appear in all benefit maps. These benefits tend to lead to the final benefits and indicate that progress is being made toward the longer-term goal. In some instances the intermediate benefits may be necessary before a later benefit can be realized; for example, increased hotel accommodations (potentially an intermediate benefit) may be needed before more tourists will be attracted to the city. Following the theme of the "vibrant city," these intermediate benefits may include:
  - Increased hotel accommodations
  - Reduction in petty crime in the city
  - Increased diversity in the types of retail and food outlets within the city
- Outcomes—The outcomes are the results of the changes made in the operational environment. Outcomes are often expressed in terms which make them difficult, or impossible, to measure. This does not diminish their importance. Outcomes are often important to explaining the story of the change and gaining support for it, and without achieving the changed operational environment, the benefits will not be realized. Examples of outcomes are
  - An active local business community attracting other businesses to the city center
  - An operational policy and events booking and management system
  - A safer and friendlier environment
- Capability—The capability represents a collection of project outputs (often from several projects) which are required to be completed and operational before the outcomes can be achieved. For a "safer and friendlier environment" to be achieved (at least in part), it may be necessary to create new pedestrian-only areas, new policing strategies, cleaner and brighter streets, signage, seating areas, access for visitors with disabilities, outdoor venues, and temporary services. Some of these will be delivered by projects, while other outputs may be developed as part of existing operational processes or change management activities; regardless of the source, time and effort that will be spent to create the new outputs. Change management activities are undertaken outside of projects and within the operational environment, either as preparation for, or as part of, the transition.

- Outputs—The products created by the projects, some of which have been listed above.
- Projects—The projects are identified, selected, and designed because they can deliver the outputs, which provide the capability to generate the outcomes that can be measured as benefits, which are linked to the corporate objectives.

The projects are the last element determined. This ensures that the work committed and undertaken is focused on delivering the benefits identified and satisfying the expectations of the stakeholders. Projects are selected based on the desired benefits they can contribute toward realizing.

A variation on this technique may be used for emergent programs. In an emergent program, some of the projects may have been identified, and some may be underway, before any benefit planning takes place. In these circumstances, it is advantageous to complete a benefits map referring to decisions made and projects which are underway as constraints at the start of the current program. This will enable the team to present their understanding of the current, and forecast, positions, to obtain the views of stakeholders and establish their expectations from the melding of the projects into a single program. There will be obvious differences between the intention of the initiative and the direction the program has already taken. The implementation of changes to the initiative needs to be negotiated with the project delivery teams and the stakeholders to reach a compromise which achieves the best results. This clarity may result in canceling some of the projects, or making significant changes that are agreed to by the stakeholders or the portfolio group, to avoid duplication of activities and effort or the application of resources to activities and projects which to not lead to the desired benefits and objectives.

Figure 6.3 is an example of a benefits map which shows how each of the projects, listed on the left-hand side of the map, is linked to the objectives on the right-hand side of the map. There is a clear chain of dependencies between the investments in the projects and the final benefits and objectives.

Workshops with the key stakeholders are often a useful approach to beginning the process of compiling the benefits map. Initially, these should be relatively short, one- to two-hour events with the key stakeholders and the program manager to agree on the corporate objective(s) and how the primary, long-term benefits which are linked to those objectives can be achieved. More detailed workshops should be held with the program manager and the participating business change managers to extend the map "backwards" by detailing the intermediate benefits and identifying any changes which will be required to the operational environment, and ultimately to identify the required outputs from projects.

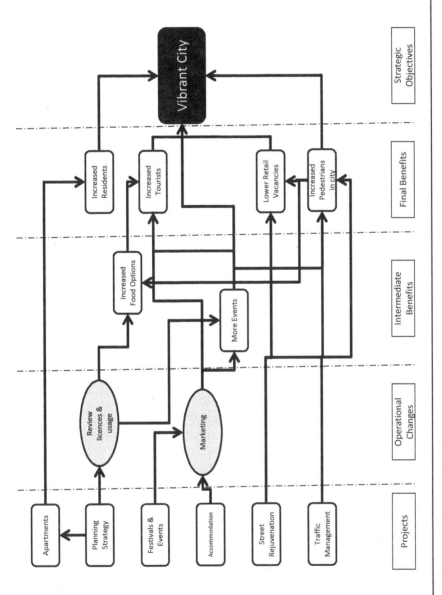

Figure 6.3 Example of a Benefits Map

A number of similar diagrammatic techniques can also be used by the stakeholders. Although each technique addresses similar issues—that is, the identification of benefits and the establishment of the relationships between the different elements—there are some key differences among them. It is generally advised that one method be selected, which the team or organization can become familiar with and use regularly. All of the methods produce diagrams which clearly show the links, or dependencies, between the technical work and the benefits and ultimately the organizational investment objectives. This clear connection between the work to be performed and the objectives can be a persuasive tool, which should assist with the engagement of stakeholders.

6.3.2 Benefits Dependency Network

The benefits dependency network (BDN), developed by Ward and Daniel (2006), uses five types of objects to show the pathway from an initiative to long-term goals:

- IS/IT enablers
- Enabling changes
- Business changes
- Business benefits
- Investment objectives

Designed initially for IT projects and initiatives, the technique is applicable in other environments and results in a diagram similar to the benefits map as shown below in Figure 6.4.

This network is constructed in a similar manner to the benefits map by starting with the end in mind. The investment objectives are stated and then deconstructed into specific benefits. One of the advantages of this technique is that the key stakeholders and sponsors can contribute toward the collection of the investment objectives, perhaps through workshops, and come to an agreement without needing to become involved in the more detailed design of the network from a technical perspective.

Once the business benefits are identified, it is possible to select the IS/IT enablers, which are required and work from left to right constructing the network to connect with the relevant benefits. The IS/IT enablers can be seen as the same as the project outputs from the benefits map. The enabling changes are the outcomes, the results of the implemented IS/IT enablers. Between these enabling changes and the business benefits are the required business changes. These are changes within the operational environment that will be required to

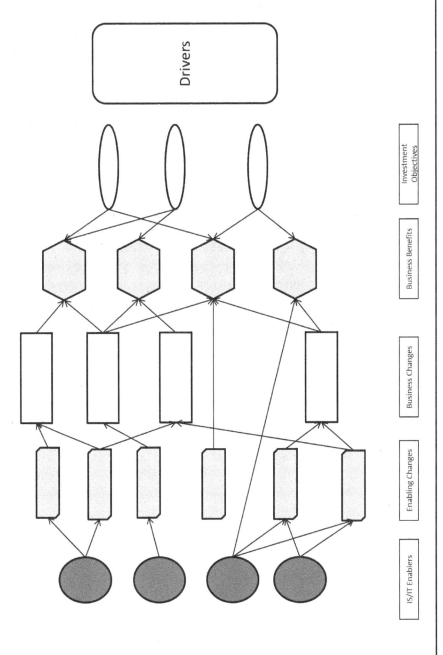

Figure 6.4 Example of the Structure of a Benefits Dependency Network

modify the behaviors within the organization, which will produce the benefits. Some of these business changes will be relatively quick to embed, and some will take significantly longer. It may be helpful, then, to categorize the business changes to set expectations among the interested parties (Bradley [2006] identified the first three categories):

- Culture—This is likely to be longer-term in nature. It takes a long time to change and embed a new culture. In many situations it may not be possible to achieve a change of this magnitude—consider the number of unsuccessful mergers between corporations of differing organizational culture. For this reason it is wise to identify this business change as long-term and to identify the risks and potential dis-benefits associated with it.
- Strategic—Another longer-term category, which will require agreement with the new strategy or policy and dissemination of it throughout the organization and its clients. Generally, there is more control over strategic change than cultural ones, but there are still a number of risks and dis-benefits, which need to be accounted for in the process.
- Procedural—Business processes and procedures may need to be changed. This is a relatively short-term category of business change. In most cases the speed of change and the timing of it will be within the control of the performaing organization and, as such, will be less contentious than the others.
- Technical—As a result of the projects and enabling changes, changes may need to be made within the business to accommodate the implementation of different technologies and a subsequent need for training, new operational and maintenance protocols, new contractual arrangements, and new relationships to be developed.
- Contractual and legal—Some business changes arise out of a new contractual relationship, for example, with a new supplier, a joint venture partner, or from changes in legislation or governing regulations. Some of these changes may be connected to other categories—for example, establishing a joint venture will involve contractual obligations and the establishment of a new collaborative culture.

6.3.3 Benefits Dependency Map

Bradley (2010) developed a variation of these models which begins by establishing the causal relationships between the benefits. This starting diagram is a network of the benefits only. The benefits dependency map then adds the following elements:

- Bounding objective
  - Measurable end goals which support the vision of what is being attempted
- End benefit
- Intermediate benefit
- Business change
- Enabler

This map is similar in appearance to the benefits map discussed previously. The benefits dependency map is normally drawn in a more structured manner, and the bounding objective is a measurable target.

6.3.4 Benefits Logic Map

The benefits logic map method was developed from investment logic mapping, which is a methodology designed and used by The Department of Treasury & Finance of the State Government of Victoria, Australia. The benefits logic map identifies five elements, often gathered into three groups:

- Problem, defined by the
  - Strategic drivers
  - Investment objectives
- Benefits, detailed as
  - End benefits
- Solution, being the projects and the short-term benefits
  - Intermediate benefits
  - Solution/initiatives (the projects)

The benefits are shown in the middle of the map, as in Figure 6.5, with the problem being defined from the left-hand side and the solution from the right. One of the advantages of this method is the separation of the problem and solution, which allows different stakeholders to contribute where they have greatest understanding and interest. The key stakeholders and sponsors define the drivers and investment objectives and identify some of the end benefits. With this information, the project delivery team and operational subject-matter experts will be able to identify the projects and intermediate benefits.

This method can easily be tailored to suit the organization, incorporating the terminology in use and familiar to the participants in the initiative. It is also clear who among the stakeholders will be involved with identifying each of the elements. However, there is a temptation to oversimplify the environment and group some benefits together. This may be an efficient way to present the

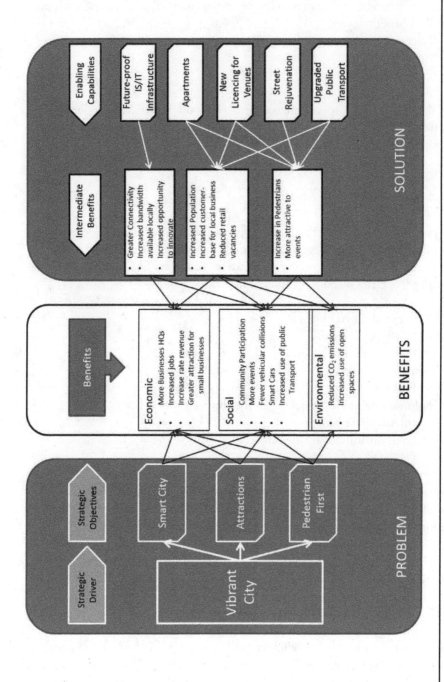

Figure 6.5 Example of a Benefits Logic Map

Table 6.1 Comparison of Diagrammatic Techniques

Technique	Advantages	Disadvantages
Benefits map	• Clear image of dependencies • Highlights the projects and operational changes	• Becomes complicated as the number of benefits increases • Needs to work back from objectives
Benefits dependency map	• Highlights the deliverables and operational changes • Clear image of dependencies	• IS/IT-focused (can be adjusted)
Benefits dependency network	• Begins with the benefits only • Establishes the dependencies between benefits	• Becomes complicated as the number of benefits increases • May identify several low-value benefits
Benefits logic map	• Places benefits in the center of the map • Clear high-level view • Clearly shows links between strategic objectives and benefits	• Groups benefits together

information, but it removes some of the dependencies between the elements, and it is not always clear which element is contributing to which other elements. One of the reasons for applying these diagrammatic techniques is to determine and highlight the causal relationships between the elements, which means that the impact of failure to meet each of the targets can be assessed and controlled. This method can sometimes obscure the true relationships between the elements if the groupings are too robust, and as a result, decisions are made based on an incomplete understanding of the network and its relationships.

Table 6.1 highlights the major advantages and disadvantages of the techniques discussed. This may enable the practitioner to select a preferred technique based on the circumstances of an initiative.

6.3.5 Applying These Methods

Developing the maps, using any of the techniques discussed earlier (or others), requires input from a number of stakeholders and ultimately needs to be approved/bought into by the key stakeholders. It is advantageous to arrange a workshop or, more likely, a series of workshops to develop the maps. This will allow the larger team to spend time analyzing the problem to be solved and to discuss it more broadly.

For smaller projects it may not be necessary to take such a formal approach. The ideas for changes and the identification of potential benefits may be generated from the operational teams. These individuals, teams, or their managers may see the opportunity to generate benefits based on their operational work and familiarity with the current environment and practices.

6.4 Identifying the Right Benefits

"The things that get measured are those which improve" is an old adage of management theory and practice. It has been attributed to many people, but I first encountered the phrase attributed to Peter Drucker and have since seen it linked to Tom Peters, Robin Sharma, and Lord Kelvin. Regardless of the source, it is relevant and important in the context of benefits realization management: focusing on the things which will be measured and committing effort and resources to achieving the targets set for them. When selecting the primary benefits and explaining them to a larger audience, the key stakeholders and the program manager must be comfortable that they have selected the benefits because they are closely aligned with the corporate objectives and will become the focus of behavioral and operational changes.

I am a fan of *Freakonomics* (Levitt and Dubner, 2005) and the other writings of Steven Levitt and Stephen Dubner. In *When to Rob a Bank* (Levitt and Dubner, 2016), which is a collection of blogs and other discussions, the authors recall a time when setting an incentive aimed at encouraging one behavior actually resulted in a completely different behavior. Levitt and Dubner (2016) noted that their Twitter followers numbered almost 400,000. They announced that they would present some merchandise to the person who signed up as the 400,000th follower, and they watched as people registered as followers. When they looked at the detailed statistics, they found that their total number of followers had actually decreased. Some forensic analysis followed, which uncovered the series of events following the announcement. There had been a rush of people becoming followers, but this was predominantly because existing followers had "de-followed" in order to re-register as a follower in a bid to win the prizes. This is a commonly observed phenomenon, whereby the selection, and promotion, of the "wrong" benefit, target, or incentive may result in responses and behaviors which are unexpected and contrary to the intentions of the initiative.

"Management is doing things right. Leadership is doing the right things" (Drucker, 2001).* This leads to the conclusion that one of the crucial roles of the

---

* Similar quotes are found in other texts, including: "Managers are people who do things right and leaders are people who do the right thing. Both roles are crucial, and they differ profoundly. I often observe people in top positions doing the wrong things well" (Bennis and Nanus, 1985).

key stakeholders of a program is to ensure that the "right things" are the focus of the team: The right benefits have been identified.

Identifying and emphasizing the right benefits are particularly important when there are multiple options which can be taken to achieve the goal. Consider productivity. Productivity is the ratio of the value of the outputs created compared to the cost of creation:

$$\text{Productivity} = (\text{Value of Outputs})/\text{Costs}$$

Any initiative aimed at improving productivity could address either aspect or both, including:

- Reducing costs while maintaining outputs at the current level
- Increasing outputs while maintaining costs at the current level
- Increasing costs and increasing outputs by a greater proportion than the cost increase

In this situation it is important to understand the context and the preferred approach to achieving the goals. Reducing costs may be accomplished by the reduction of maintenance and servicing of equipment or the use of less expensive raw materials. Both of these examples will certainly reduce costs in the short term but may lead to longer-term problems when the plant fails or the less expensive, and presumably lower-quality, materials lead to poor products, resulting in unhappy customers. The key stakeholders in the sponsoring group should steer the team toward actions which will fit with the benefits and the intention of the initiative. The clear identification of the relevant benefits will ensure that the teams undertaking the initiative remain focused on actions which will lead to these benefits.

The following case study demonstrates the importance of understanding the benefits and the changes in the behaviors of the stakeholders which will be required to reach these goals.

A police force identified a number of intersections as high-event sites—locations where a high number of collisions and near-misses occurred. The department wanted to take action to reduce the risks and dangers at these sites. Cameras were installed at these locations to capture speeding offenses and other traffic offenses. After a period of use, the results were reviewed.

The process followed when a vehicle went through a red light or was speeding (triggering a photograph and registering the offense) was as follows:

1. Photographs were taken and transmitted to headquarters.
2. Photographs were reviewed by the Traffic Division, and the car registration details and time of the incident were verified. These actions registered the initial offense.

3. Photographs were passed on to another team to check for other infringements, such as not using seat belts or damaged lights, etc.
4. All infringements were compiled into one notification, and the letter of infringement was sent out to the registered owner of the vehicle.

The results after a period of operations were surprising. The volume of notices issued increased (and because each resulted in a fine, revenue increased), but the number of offenders stayed constant. The increase in fines was due to the registering of all offenses 24/7 and the identification of the secondary offenses through step 3 of the process. As a result of steps 3 and 4, there was a lag between the offense, the notification of the infringement, and the fine. This led the offenders to separate the two events—the gap was too great, so they did not associate the consequence with the offence. As a result, there was little change in driver behavior, and therefore, no reduction in the number of offenses or events at these locations.

The primary benefit sought was a reduction in collisions and dangerous events at these known problem sites. This failed initially because business changes were not made to the internal processes, which would have supported the overall objectives. Armed with this information, the department decided to make business changes, which included separating the notice of offenses so that an offender would immediately be informed of the primary offense (running the red light and/or speeding), and the secondary offenses would be dealt with separately and later. This reduced the gap between the offense and the notice of the fine significantly. Once this lag was consistently reduced, from over 30 days to less than 14 days, the number of offenses suddenly and significantly dropped.

So, what was the "right benefit"? Number of offenses? Volume of fines? Number of days between event and notice? The answer depends on the context and the primary objectives of the initiative. In this case, it is also important to note that a well-executed project can yield only a few benefits if the associated business changes are not implemented successfully.

6.5 Who Identifies the Benefits?

It seems an innocuous question! But it is so important to the results. It would be too simplistic to declare that everyone should be involved in identifying the benefits, but this is true, to a degree.

It may be easier to discuss the possible limitations of the main protagonists in the initiative. The Sponsor and other key stakeholders can be expected to identify the primary objectives of their investment and with them the key measurable performance indicators (benefits). However, there may be a significant number of other benefits, which might contribute strongly to the business case and the success of the initiative, of which these stakeholders are not familiar.

The program manager and team may be involved in identifying the benefits. However, this group includes the specialists in delivering the technical solution and not necessarily experts in the operational use of finished products. For example, the construction team knows how to build the bridges, roads, and tunnels and are not necessarily specialists in the field of traffic management. The team who will deliver the technical solution may be a contractor, or a group of contracted companies and personnel with limited knowledge of the operating environment. They will certainly have little or no authority to make changes to that environment. So, the ability of this team and its personnel to identify benefits may be limited.

In summary, there needs to be input from a cross section of the stakeholders when identifying the benefits. This includes the senior stakeholders and investors, the delivery team including the program or project manager and their teams, and the operational personnel who can shed light on the results which can be expected to be achieved, and the changes in the operational environment which will be required to achieve the benefits.

As stated earlier, benefits are generated within the operations of the organization. Therefore, it follows that a major stakeholder in any benefits realization environment will be Operations and their personnel and teams. Not only will the operational teams be helpful in identifying the benefits, they should be invited to contribute to:

- Quantifying the benefits
- Determining the ease of realization
- Identifying risks and opportunities
- Using lessons learned from the operations
- Promoting a pragmatic view of the benefits which will be created/generated
- Determining the connections and dependencies between the projects and benefits
- Identifying the business changes required as a result of the projects

Senior stakeholders and "investors" should be involved in identifying the benefits so that they can shape and take ownership of the direction of the initiative. It is important that these stakeholders understand, and agree on, the benefits, as these will be the tangible and measurable manifestations of their investments. They will be able to prioritize the benefits they expect and in this way will be in a position to determine the results they associate with success.

The project delivery team should contribute to the process of identifying the benefits. If they are unfamiliar with the operational environment, it will be important that the project delivery team understand the perspective of the users and their criteria for success. This will include gaining better knowledge of the performance criteria, which will be of interest to the users and will be connected to the benefits. The project delivery teams are charged with developing and delivering products that can generate the benefits when they are put into the operational environment. The greater the teams' understanding of the benefits, the more likely the teams will deliver something which will be acceptable and useful.

6.6 Documentation

6.6.1 Benefit Profile

Using the mapping techniques leads to the use of brief descriptors of the benefits because the benefits names/titles need to fit within the box of the diagram. This can lead to stakeholders holding differing interpretations of the definition of the benefits. To address this concern, a detailed description of each benefit, called a benefit profile, can be developed. Just as each element of a project is given a specification, or product description, to ensure a common understanding of the characteristics and attributes of the component, each benefit should have a similar definition, which is agreed up and accepted.

The benefit profile contains the detailed information concerning the benefit, including:

- Description of the benefit
- Projects on which the benefit is dependent
- Business changes required to be implemented before the benefit may be realized
- Time between the project completion and the benefit being realized
- Method for measuring the benefit and who will be responsible for that measurement
- Baseline measure and supporting evidence

Having an agreed-on benefit profile avoids any disputes later in the benefit realization pathway and provides a clear assessment of the definition of the benefit.

Experience has shown that the formalization of the definition of each benefit instills sound practices within the organization and encourages a healthy attitude toward the discussion of benefits.

Anecdotally, it has been found that the discipline of writing benefit profiles encourages members of the organization to present ideas for change and causes the team to consider the method of measuring both the baseline and future benefits earlier. One of the most significant advantages to producing benefit profiles is the discipline which they bring to the discussion of scale and measurement; for example, fewer benefits are defined in terms which are intangible, that is, the process is "better."

Almost as important as these points is the noticeable change in attitudes within the organization, which accompanies the benefit profile. Technical team members talk more about what they will achieve rather than what they will do. Teams and individuals see and discuss the value that their work can add to the organization more than before. Teams conduct different discussions with the stakeholders based on added value rather than functionality. Stakeholders' interests become the focal point of the projects, and the outcomes and benefits gain a high profile as the fundamental reasons for investing.

6.6.2 Benefits Register

A benefits register is a summary of the benefit profiles. It is a list of the benefits with details of the attributes of each one. There may be some duplication between the contents of the benefits profile and the benefits register. This should be accepted, and any effort to combine the two into a single document should be avoided. Through the discipline of writing, and thinking deeply about each benefit, each benefit profile is important in gaining a clear understanding of the work involved, and potential risks associated with the generation of each benefit is important in gaining acceptance of benefits realization management.

The benefits register should contain the following:

- Reference number for the benefit
- Benefit title
- Size of benefit
- Reference to baseline metrics
- Period over which the benefit will be realized
- Dependencies
  - Project
  - Transition activities
  - Other benefits
- Benefit owner
- Date identified

Table 6.2 Activities Conducted during the "Identify the Benefits" Process

Activity	Purpose	Responsibility	Documentation
Workshops with the sponsoring group	Agree on objectives and final benefits	Program manager	
Workshops with the program team and the business change managers (BCMs)	Identify the benefits and outcomes	Program manager	
Identify the benefits	Compile the benefit profiles	Program manager and BCMs	Benefit profiles
Map the benefits	Understand the dependencies between benefits and identify the projects	Program manager and BCMs	Benefits map (or similar diagram)
Compile the benefits register	List all of the benefits and dis-benefits which have been identified	Program manager	Benefits register

- Date realized
- Measurement approach
- Notes

6.6.3 Benefits Map (or Other Diagrammatic Representation)

The benefits map, or another diagram, is a single image showing the dependencies between the projects and their benefits. It should always be accompanied by additional information to ensure that the complete "story" is told.

The image should contain:

- Benefits and dis-benefits
- Outputs
- Outcomes
- Strategic objectives
- Dependencies within the network

Additional notes regarding the following should be compiled:

- External dependencies
- Major risks to the realization of benefits
- Comments regarding the sources of the information compiled
- Business changes that may be required for the sustainability of the benefits

Table 6.2 summarizes the activities conducted during the "Identify the Benefits" process.

6.7 Summary

Identifying the benefits is an important and exciting part of the process. It allows and encourages interaction among the stakeholders, sponsors, the project delivery team, and the operational teams. This identification process inevitably takes time, and resources must be allocated to undertake this valuable work.

The use of diagrammatic and mapping methods should help to achieve a better understanding of a complex situation and allow for decisions to be made with a clear understanding of the influence and impact these decisions will have on the overall system. A number of methods have been discussed in this chapter, and there are also other proprietary methods which can be used. It is important, for learning lessons and increasing organizational maturity, that one method be selected and persevered with, to gain familiarity with its use. This increased familiarity will lead to a greater understanding of the complexity of the changes which will follow.

The next chapter will discuss methods for quantifying and prioritizing the benefits.

Some of the issues addressed at this stage should be

- Have the investors been included in the identification of benefits?
- Have too many benefits and dis-benefits been identified?
- Have baselines been established for each of the benefits and dis-benefits?
- Have the operational teams been included in the process?
- Is there a natural flow from project-outcome-benefit-strategic objective?
- Have projects been selected for the contribution they make to the benefits network or for some other reason?
- Have detailed and approved benefit profiles been developed?

Exercises and Activities

Use the following examples of final benefits from an infrastructure project and an IT initiative to develop a benefits map (or similar) to identify the short-term, intermediate benefits, changes, and projects which could be conducted.

1. *Road and Public Transport Program*

Problem: Traffic from the northern and western suburbs converge to one junction, which has reached capacity and resulted in congestion at that point. Travel times into and out of the city have lengthened for car passengers and bus traffic. In addition, feeder roads from the city become gridlocked during rush-hour periods. There is no capacity to build additional roads because of the location of the business district and the protected "green belt areas." A "green belt" is an area of open land, often parklands or woods, surrounding a city or suburb which is intended to be left undeveloped and provide some natural environment to limit the urbanization of the area.

The following long-term (final) benefits have been identified as contributing to the objective of an "accessible city center":

- Improved travel time and reliability for existing bus services
- Reduced traffic congestion at the northwest corner of the city center
- Improvements to traffic safety
- Improvements to pedestrian and cycling access and safety
- Reduced congestion on feeder roads

2. *IT Program*

Problem: A city council, with a number of teams of inspectors, has several teams who must be mobile in order to gather data from the recreational areas, grass and wooded areas of council lands, and also for development locations (both residential and industrial). The current system requires writing notes and support staff entering the data into the current internal database.

The following long-term (final) benefits have been identified as contributing the objective of "accessible data for all" and "productivity improvements":

- Reduction in handling a lot of paper
- Reduction costs and time associated with retyping from filled-in forms
- Reduction in report delays while waiting for someone to update
- No errors in re-entry (data quality improvements)
- Simplified entries through popup lists—reduction in number of categories and terms
- Faster information access.

Chapter 7
Assess the Benefits

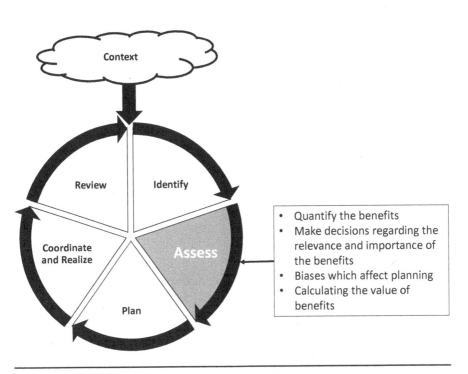

Figure 7.1 Assess the Benefits within the Benefits Life Cycle

> *"Anytime emotions are involved, you cannot come up with an impartial and objective assessment of any given problem."*
> – Benigno Aquino III

> *"Everything that can be counted does not necessarily count; everything that counts cannot necessarily be counted."*
> – Albert Einstein

> *"Is your want worth wanting?"*
> – Akiroq Brost

"Identify the Benefits" should be viewed as purely the action of identifying the potential measurable changes in performance (both positive and negative) which are likely to be realized as a result of completing the projects and making the operational changes. While identifying these benefits, it is tempting to become distracted into analyzing these benefits more deeply, but this should be avoided, as shown in Figure 7.1, where identification and analysis are shown in sequence. One of the major reasons for separating the identification and the more analytical activities is that the skills required for each one are quite different, and the pace at which these activities can progress is different, meaning that the people involved in the intensive identification process may not have the time or the patience for the slower-moving and more detailed analysis. It is important to engage the key stakeholders early in the project, and the identification of benefits activities is an opportunity for this group to contribute, but the project risks alienating these important supporters of the project if they are required to commit significant amounts of time to the assessment of benefits when others may be able to address this area more efficiently.

Assessing the benefits comprises two distinct activities:

1. **Quantification of the benefits**—During which an analysis of each benefit is conducted to determine the size of the benefit and the dis-benefit. This includes:
   - The baseline against which future measurements will be compared
   - The size of each benefit—its financial value or other measurable property
   - The cost of transition to the new methods of operation and maintaining the new regime
2. **Decision making regarding the relevance and significance of each benefit**—When key stakeholders confirm their priorities and make decisions regarding the formulation of the business case:
   - The likelihood of successfully realizing the benefit
   - The risks associated with the operational changes required to realize the benefits

- When the benefits will be measured, and expectations of the value at each point
- Whether the added value will be worth the effort

Once this information has been gathered and assessed, decisions can be made regarding the priorities for the benefits, and indeed, which ones will be discarded. These decisions are guided by the context of the initiatives and instructions from the sponsor.

This chapter will detail methods for quantifying benefits, and the pitfalls which face the team, and the decisions which need to be made before plans can be developed.

7.1 Quantifying

Quantifying refers to the process of estimating a value for each benefit, a value which represents the forecasted magnitude of the advantage or disadvantage. This should apply equally to financial and nonfinancial benefits.

Quantifying the scale of a benefit, or group of benefits, is difficult and should always utilize the most up-to-date performance information regarding the baseline levels and for forecasting. This initial quantification of the benefits provides an estimate of the future performance, if the initiative is successful, and will be a valuable input into the business case. As the initiative progresses, the information available to the team and stakeholders will change, and adjustments to these forecasts may be necessary. It is important that the forecast of benefits is reviewed as the program progresses, and as new information becomes available, which may support or contradict the earlier figures. These reviews can be conducted on a regular basis, such as quarterly, or at logical points during the initiative, such as at the end of a project or transition period.

7.1.1 Cognitive Bias—Some of the Traps

A number of factors always influence the approach to estimating and planning. These apply to project planning and the quantification of benefits equally, and if unchecked, they will skew the accuracy of estimates and the impact on the quality of decision making.

Cognitive bias refers to a tendency to deviate from sound or rational judgment, resulting in drawing illogical or prejudiced conclusions about a situation. A large number of these biases have been identified (on the order of 100), including, ironically, the bias blind spot, which is the tendency to see oneself as less biased than other people and to be able to identify more cognitive biases in others than in oneself. This leaves sufferers with the belief that they are correct because they do not fall into the cognitive bias traps while everyone else involved does—resulting in the sufferers ignoring the advice of the others. Other cognitive biases include:

- Base rate fallacy—The tendency to ignore base rate information (generic, general information) and focus on specific information (information pertaining only to a certain case) (Baron, 1994).
- Ben Franklin effect—People who have performed a favor for someone are more likely to do another favor for that person than they would be if they had received a favor from that person.
- Courtesy bias—The tendency to give an opinion that is more socially correct than one's true opinion, to avoid offending anyone (Ciccarelli and White, 2014).
- Dunning–Kruger effect—The tendency for unskilled individuals to overestimate their own ability and the tendency for experts to underestimate their own ability (Kruger and Dunning, 1999).

The following discussion covers some of the common biases which affect planning and forecasting in particular, and goes on to how they affect the forecasting in a project environment and the quantification of benefits in particular.

Awareness of these cognitive biases, and their ability to impact estimating and forecasting activities, empowers the team to search for accurate and reliable sources on which their plans can be based. By recognizing these influences, it is hoped that responses can be enacted which address and remove the influence of the biases, making the resulting metrics more accurate.

Strategic Bias

Within the project environment there has been a history of inaccurate estimating, especially for large-scale projects. This applies to the estimating involved in planning and forecasting the value of the investment, that is, costs, schedule, and risks, as well as estimating the benefits. Considering the availability of experienced estimators and schedulers, and data which could be used to validate the estimates, it does not take a conspiracy theorist to make the leap that the estimates may be deliberately adjusted to be more palatable to the sponsors. Hirschman (1967) concluded that if people knew in advance the real costs and challenges involved in delivering a large project, few projects would ever start. Ignorance may be a crucial element in getting initiatives approved! This leads to the deliberate and strategic approach to underestimate the costs and duration of the work necessary, and the exaggeration of the benefits. The initiative can then start in a wave of optimistic "realism"; once started, it will be difficult to stop because of the sunk-costs effect and embarrassment at admitting the initial mistake or a poor decision. Flyvbjerg (2014) cited Willie Brown, former mayor of San Francisco, commenting on the San Francisco Transbay Terminal megaproject:

> News that the Transbay Terminal is something like $300 million over budget should not come as a shock to anyone. We always knew the initial estimate was way under the real cost. Just like we never had a real cost for [the San Francisco] Central Subway or the [San Francisco-Oakland] Bay Bridge or any other massive construction project. So get off it. In the world of civic projects, the first budget is really just a down payment. If people knew the real cost from the start, nothing would ever be approved. Start digging a hole and make it so big, there's no alternative to coming up with the money to fill it in. (p. 12)

To start any initiative under such terms will lead to a large metaphorical hole being dug too. Where will the necessary and unbudgeted money be found? What will suffer in the future to meet the needs of the underestimated project? What is the likelihood of realizing sufficient benefits to cover the inflated costs?

The deliberate decision to play a strategic game to obtain the approval for an initiative may take the form of underestimating the scale of the investment, downplaying the risks involved, overestimating the benefits, and using popular "buzz-words" which are favored by the stakeholders. The use of "buzz-words" to gain support for projects is commonplace; it involves the inclusion of seductive and attractive arguments based on the language favored by the organization. IT projects which "increase mobility," or projects which use new technology to reinforce the organization's cutting-edge image, can be made to appear exciting and attractive, even when they are not viable. Linking the project to the strategic plans of the sponsoring groups is a strong way of showing the "value" of the investment. This often does not require quantification, because stakeholders may be swayed by the argument of linking action with the strategic goals of the greater organization(s). Convincing emotive arguments take the place of accurate analysis and planning.

To deliberately mislead investors and sponsors by presenting deflated costs and inflated benefits for a project could create legal and ethical issues, potentially resulting in disciplinary action and prosecution. If this practice was widespread, there would be many more headlines and cases regarding this malpractice. However, it should be possible to demonstrate enough diligence to present a defendable position which represents an attractive perspective to these stakeholders. This can be done by demonstrating that data and subject-matter experts were sought for their input into the process.

This issue is open to influence from some of the other biases, which will be discussed shortly. For example, using information which already exists from previous projects (which most would consider a good approach), estimators could conclude that a new project should cost $1 million. If this confirms a prior assessment (for example, the sponsor's original guesstimate) of the project, the estimators may not look for further corroboration. These data were available and confirmed an assumption—making it susceptible to two biases (availability bias and confirmation bias). This position is defendable, and it would be difficult to argue that a reasonable estimate was not produced, but it is possible to have calculated a flawed estimate because of the selective nature in which data were used.

Optimism Bias

Optimism is a double-edged sword with respect to the realization of benefits. Optimism bias is the tendency to be overly optimistic and focus on outcomes which are favorable. As such, it is closely associated with the planning fallacy and illusion of control. In many ways, this can highlight the danger of several biases acting simultaneously and reinforcing an unrealistic belief.

Optimism bias may lead the stakeholders to underestimate the cost and time for delivering projects because they are confident that the work will go well and that there will be few problems. Threats, or risks, are underplayed and overlooked. This can reinforce the planning fallacy, which stipulates that for this project, "Everything Will Go According to the Plan" (EGAP). This may be contrary to lessons from other, similar projects and creates an overconfidence which is not based on evidence.

On the other side of the business case equation are the benefits. Optimism bias is likely to cause the team to exaggerate the forecast benefits to show a more appealing result. This leads to a business case based on underestimated costs and overestimated benefits with risks downplayed, making the investment look attractive when it is fraught with danger.

Elaurant and McDougall (2014) elaborated on the results of Flyvbjerg (2003) in confirming that the costs for megaprojects typically overran official estimates by 50% to 100%, and user revenue or patronage (benefits) was usually 20% to 70% less than forecast. Unless the benefit–cost ratio is very high, this optimistic approach to planning and forecasting will cause catastrophic failure.

Merrow (2011) supported this gloomy outlook of industrial megaprojects, estimating that 25–50% cost overruns for liquefied natural gas (LNG) projects are common. Furthermore, of the LNG megaprojects which failed, the average production was 41% of the forecast. In the case of LNG projects, production equates to benefits.

The Planning Fallacy

The planning fallacy leads the team to believe that this project will go according to the plan. Regardless of previous experiences and evidence that the tasks are difficult and have resulted in poor outcomes in the past, the planning fallacy creates a sense of optimism that on this occasion the bad luck that has beset previous similar attempts will be offset by the good planning.

This creates an attitude of invincibility among the team, who will believe that nothing can go wrong—this time. The team will credit any success to the planning and management capability and the brilliance of the team and blame any failure on pure bad luck.

The planning fallacy can be dangerous to the team's success because complacency can set in later in the project and post-project. The team takes the accuracy of the plan for granted and can imagine no other approach or event (risk) which might derail the progress. The belief that "Everything Will Go According to the Plan" (EGAP) prevails.

In his paper "Planning Fallacy or Hiding Hand: Which Is the Better Explanation," Flyvbjerg (2018) discusses this phenomenon. Some will argue that there is a guiding "hand" which uses experience to "magically" solve the problems that arise, making it seem that there is a benevolent guide or a helping hand. This point is advanced by human ingenuity, which often comes to the rescue during projects and is expected to arrive in times of need. The experience and ingenuity of the team often rescues the project from overly optimistic and unrealistic plans.

On the other hand, there may be a dark side to the force, one which conjures up all manners of risk and complexity and hurls this at the team from time to time. The uncertain nature of programs and the resistance to change can create hostile and indifferent stakeholders.

Group Dynamics

Particularly during estimating activities, project teams are encouraged to collaborate with stakeholders to seek their input and share experiences. This is a valuable opportunity to engage closely with some of the stakeholders and to gather information from a wider set of experiences. However, groups can often produce some unexpected results and are influenced by three important factors.

1. **The bandwagon effect.** The term *bandwagon effect* is derived from the political environment, in which voters may be influenced by the changing behaviors of others. As one politician becomes more popular, more voters are swayed to join his or her cause and to vote for that politician—hence the term "jumping on the bandwagon." The term is used broadly to highlight the fact that people change their position and opinions to conform to those which are gaining popularity. McAllister and Studlar (1991) discussed this phenomenon, and Mehrabian (1998) demonstrated this effect during studies of political campaigns.

   The term originated when politician Dan Rice used a "bandwagon" during his public appearances. As his popularity increased, other politicians literally jumped on their own bandwagons to improve their visibility and raise their profile, and (hopefully), popularity. The term is often used in sporting situations, where people are drawn to support an athlete or team as their success increases. In a project context, the effect is for people to be drawn into agreement with a belief which has gained popularity among the group members. This obviously makes it more difficult for one person to express an opposing or less popular opinion.

2. **Groupthink.** Groupthink is the situation whereby there is tendency for individuals within a group to moderate their views toward a "norm." It creates an Orwellian* situation where there is little or no dissent, nor is it perceived to be polite to raise views which are not widely shared within the group. The group tends to confuse fact with assumptions because of the lack of opposition, dissent, or open discussion within the group. This tendency is reinforced when the majority of those involved share the same set of beliefs and values. This is often observed when the group contains members who come from a similar background—that is, the group may have a common technical background such as engineering, or they may be long-term employees with the same organization.

3. **The Abilene Paradox.** The Abilene paradox occurs when the outcome of a group's actions contradicts the intent of the decision. The name is derived from an anecdote told by Jerry Harvey (Harvey, 1974):

   > A woman and her husband are spending time with her parents in Coleman, Texas. It is a hot day, and the woman's father suggests that they take a trip to Abilene (about 50 miles away) for a meal. The others are not sure but agree to "go with the flow" rather than rejecting

---

* George Orwell published his dystopian novel *Nineteen Eighty-Four* (in 1949), which foretold of a domineering ruling political class which punished free thinking and encouraged conformity to the official version of the "truth."

the idea. The trip, in a car lacking air conditioning, is very uncomfortable, and the group returns to Coleman after a lunch that was not particularly good. On the return journey, all of the group whine and complain that this was a bad idea and a poor decision. The group are puzzled that they have just spent time and resources doing something that none of them really wanted to do, and the proposer (the woman's father) states that he only proposed the idea because he thought the others might be bored.

How did the group agree to do something that none of the participants wanted to do? The phenomenon is explained by theories of social conformity and social influence, which suggest that humans are often very averse to acting contrary to the trend of a group. The Abilene paradox is exhibited by group members acting, or agreeing to act, in a manner which contradicts their desires or beliefs—making it different from Groupthink. Harvey (1988) cited the Watergate scandal as an example of the paradox in action.

These three factors result in individuals conforming to a pack mentality and reaching agreement, and consensus, which are heavily influenced by the other group members. These effects are amplified if there is a power imbalance in the group. These results are caused by poor management, or ignorance of the group dynamics.

Confirmation Bias

Confirmation bias is a tendency to seek, or rely on, information which supports the initial assumptions. This can be achieved by searching specifically for such information or discarding information which opposes the existing beliefs and assumptions. As humans, we are inclined to pay more attention to, or value more highly, information which supports our expectations. This is a case of the self-fulfilling prophecy, where there is an expectation of a particular result and evidence is sought to reinforce our assumptions, often to the exclusion of other facts. Information which does not support these expectations is discarded as an anomaly, an outlier, or is discredited. This bias creates overconfidence in the values estimated or forecast because there is evidence in support of them, even if that evidence is selective. Confirmation bias may be found in three forms:

1. **Biased search for evidence.** When a team has made assumptions or developed a hypothesis, they tend to test these in a one-sided way. Questions

are constructed in a manner that seeks agreement or confirmation of the assumption. By searching and questioning with an approach designed to confirm the beliefs, the investigation is conducted in a biased manner.
2. **Biased interpretation of evidence.** In this instance the evidence has been found, and it is the manner in which analysis and interpretation take place which exposes the bias. This results in the team giving greater credence to information which supports their position and discrediting or discarding information which does not fit the assumptions. Although all of the data and information may be available, they are not all judged equally.
3. **Biased memory.** People can often suffer from "selective recall" or "confirmatory memory." When it is known that an outcome or attribute is valued, our memory tends to recall those characteristics more easily. For example, if the project team knows that one benefit is favored above the others, the team and stakeholders will more quickly and readily remember how the project will contribute toward that benefit at the expense of others.

Availability Bias

Availability bias manifests as the tendency to place a higher value on the information which is available, or can be readily obtained, than on information which requires more research and effort to acquire. The approach is akin to saying that the more difficult information is to obtain, the less relevant that information must be. Or, put another way, "If the information was important and relevant, it would be easier to gather or obtain."

This approach leads to a lack of detailed investigation into, and consideration of, less obvious issues and ignores unpopular options and options that require more research. The stakeholders may argue that the "important" information is available, and further investigation will not clarify the situation nor assist with decision making.

The impact of this bias can be amplified when it works in tandem with others, for example:

- *With confirmation bias,* whereby information is sought which confirms the preconceived notion. This combination leads to seeking only information which reinforces one view, meaning that all of the *available* information supports that view. Unfortunately, this gives an appearance of thoroughness and completeness of the research because some effort was made to obtain information, and all the information agreed with our assumptions.
- *With groupthink,* which leads to less questioning or interrogation of the information presented before a common view is formed. This combination of availability and common interpretation can be powerful.

Loss Aversion

Losses hurt far more than an equivalent gain pleases!

Losing $1,000 hurts more than winning $1,000 provides pleasure. In fact, Hastie and Dawes (2001) found that "most empirical estimates conclude that losses are about twice as painful as gains are pleasurable" (p. 288)—making a loss of $1,000 similar in impact and significance to a stakeholder as a benefit of $2,000. This skewed approach means that dis-benefits may weigh more heavily on the stakeholders and lead them to making more conservative decisions than normal.

This is an important factor when considering the benefits which will be documented and recorded. When considering a proposal, one which emphasizes the downside of not proceeding (i.e., money or lives will be lost) will appear more compelling than one which emphasizes the advantages of going ahead (i.e., money or lives will be saved). The same benefit may be used in this case, but the presentation is more powerful in the first instance. The first situation also creates a driving force to take action—a call to action.

Kotter's Change Model (Kotter, 1996) holds that the first step in managing a change is to create a sense of urgency. This idea has been modified in Figure 7.2 to use language which is more applicable to the portfolio environment. This establishes a powerful motivator for the stakeholders and can be translated into a strong and compelling business case. Having a negative statement is often one way of getting the attention and interest of some of the stakeholders—simply because the "loss aversion" effect exaggerates the magnitude of the problem.

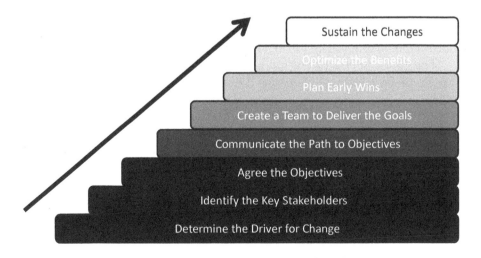

Figure 7.2 Kotter's (1996) Change Model Adapted to Portfolio Management

It is important to consider each investment which is expected to yield benefits at its core as a change initiative. In practice, something must change to achieve the results and benefits expected.

Sunk-Cost Effect

There are many situations where projects should have been stopped because they were (getting) out of control, and forecasts indicated that the objectives would not be met. One of the common arguments for continuing is that too much money has already been spent (sunk into the project), and that to stop the project would amount to throwing that money away. This effect is compounded by the loss aversion effect, which determines that any losses will be viewed as more significant than similarly sized gains.

"It is better to continue and get something out of the project than it is to abandon the project completely," is often heard. When the budgets and expenses are large, as is often the case, it is easy to be swayed by this emotive approach. However, continuation of projects for the sake of not losing the sunk costs may lead to enormous cost overruns, because the stakeholders become more concerned with saving face than recovery of the project. The objectives of the project often become clouded in the drive to rescue the project and recover something.

Generally, when discussing the sunk-cost effect, the focus is, understandably, on the financial aspects of the investment. However, there may be non-financial elements to consider, including the time and effort (which could be consolidated into costs) and reputation damage. For a change project, there will normally be some early communication regarding the change, its objectives and goals. To admit later that these are not achievable may be damaging to the organization's, or specific individuals' reputations.

The sunk-cost effect alone can be disastrous to the decision making within a project, but it becomes increasing difficult to address if it is combined with other biases, for example:

- With the planning fallacy—Money has already been spent, and we will lose it all unless we continue, *and* the remainder of the project will proceed in accordance with the plan. It will require some radical and lucky action to complete a project in accordance with a plan which has already proven to be unreliable for the first part of the delivery process.
- With optimism bias—The delivery of the project has been a struggle, *but* the benefits will be greater than we anticipated. This will ensure that a return is realized for the investment even though the progress to date has been costly.

The Endowment Effect

Generally, a higher value is placed on something we already own than we would be willing to pay for it. On the flip side of this effect, lower values are placed on something not in our possession. In the context of benefits realization management, this manifests itself as the tendency to overvalue the benefits forecast, once the initiative is underway.

For a practical example of this effect, consider the house that you own. Assume that the market value is $500,000, as derived from an independent valuation.

Imagine that you did not own that house, and you saw it advertised for sale. What would you be willing to pay for this property, if you did not already own it? Probably the first figure that was considered would be beneath the market valuation, certainly not more than that figure. Let's say that your valuation as a prospective buyer is offering $450,000.

Consider putting your house on the market for sale. What would you expect to receive for selling your home? What would you demand as a minimum sale price? Probably $600,000.

We value things we already possess more highly than things we do not own.

This valuation can have a dangerous impact on a program or project when assessing progress after commencement. The progress report may indicate that several projects are complete, and some of the benefits are beginning to be realized but not in the quantity expected. Benefit A has been measured and is as forecast, and is expected to continue to accrue as planned. Benefit B has also been measured, and the result is that only 25% of the expected benefit has been realized so far, with no likelihood of recovering a greater proportion of the forecast. The tendency will be to claim that Benefit A was the more important benefit, and Benefit B was relatively unimportant. That is the endowment effect in action. This form of bias becomes more important as the initiative progresses, and the "assess the benefits" process is revisited within the benefits life cycle.

The Affect Heuristic

If there is a known preferred option or outcome among the stakeholders, the affect heuristic presents as the tendency to be optimistic when discussing the preferred option and more pessimistic when discussing the alternatives. This often takes the form of:

- Understating the risks and issues associated with the preferred option
- Providing optimistic estimates for the costs, timeframe, and benefits of the preferred option

- Exaggerating the risks and issues associated with the nonpreferred options
- Stressing the dis-benefits and the negative elements of the nonpreferred options
- Underselling the benefits of the nonpreferred options

The affect heuristic can be subtly applied by the manner in which information is presented, especially if the primary method of communication is face to face, with little supporting documentation. This bias is not selective in the choice of information investigated in the same manner as the conformation or availability biases. However, the presentation of the information is skewed to favor the preferred option, and it exposes the decision-making process to a lack of diligence.

Regression to the Mean

More correctly, it is the ignoring of a regression to the mean which is the issue. This is a trap which is most likely to impact the monitoring and controlling of progress and expectations of the stakeholders. With enough data points, or measurements, it is expected that the cumulative results will move toward the mean, or average, result. Individual measurements may be outliers and may be unrepresentative of the ongoing levels of performance, but over time, performance (in the project and program environment that relates to the performance of delivery and realization of benefits) will revert to its norm. Care needs to be taken not to place too high an emphasis on extreme results, which may be anomalies and are unsustainable.

Consider almost any sport; athletes may get onto a "hot streak" and play well above their historically normal level. This extreme standard, along with its associated results, will not be sustained over the long term, and the athlete will revert back to his or her mean or norm. It is exciting to watch the hot streak, and, as a fan, speculate how things might look if it continues—extrapolating these outliers into the future will look impressive. But this rarely happens: Even the great athletes find sustaining their best form to be difficult.

In a similar vein, when initial progress in a project exceeds expectations, or lags behind the schedule, these extreme levels of performance are unlikely to continue for the duration of the project, and performance will normalize as the team settles into its tasks. Similarly, with benefits, there may be an initial unanticipated result, or set of results, when measuring the benefits. This may be an unexpected level, and there is a temptation to modify the forecasts for the future benefits. This should be resisted strenuously until more results confirm that the measured change is sustainable, or more likely, will regress toward average behavior. It is dangerous to make changes to the plans based on a single, or small population of data point(s) and set significantly different targets and expectations among the stakeholders.

Framing

In some instances, it is not the sources of information which create the bias, but the manner of the presentation of the options. The information or options may be presented in a way which influences the decisions and causes decisions to be made that are not optimized. Framing occurs when irrelevant options cloud the primary issues. Simonson (1989) reported this phenomenon and noted that choices will be influenced by framing, which affects the outcomes and realization of benefits.

Consider choosing between two similar laptops. Both have identical features except that the first laptop has more memory than the second option. The second option has better screen resolution. Choosing between these two options may be difficult. If a third option, which is inferior to the first laptop in both aspects, that is, less memory and lower screen resolution, is introduced as a decoy, it will cause the first two options to be viewed in a different light. Simonson found that where the third, decoy option was available, more people selected the first option.

In essence, the introduction of the decoy option made the first laptop look better because it was superior in both of the aspects which were being assessed. The reality, however, was that a preference had to be made between the two differing functions, and the introduction of the third option clouded the issues and interfered with making a choice between the two criteria which were being considered.

Anchoring

There is a danger during estimating that disproportionate credence is given to the first estimate produced. Once this initial estimate is established in the minds of an individual or the team, it is difficult to adjust this figure significantly, regardless of the evidence. If the first estimate is given by an authority figure, such as a subject-matter expert or a leader within the stakeholder group, it will be even more difficult to move away from the anchor because of the regard in which the authority figure is held. In many cases, the anchoring position is often presented in an attempt to be helpful or to set boundaries (upper or lower limits on the estimate). However, there is real or perceived pressure to confirm that this figure is reasonable.

An example of anchoring and adjustment is often used as a sales technique. A used-car salesperson (or any salesperson) will often offer a very high price as a starting point for negotiations. This figure may be well above what is considered fair value. Because the high price is an anchor, all of the bargaining is focused on reducing the price, and significant effort may be expended with little progress. The anchor makes it difficult to move a long way from that starting point; often any adjustment is seen as a success. This is true for the purchase of products and for estimating the costs and timeframe for delivery of the project and the value of its benefits.

Kahneman (2011, p. 124) demonstrated this effect in a number of experiments, including the Exploratorium study regarding environmental damage from oil tankers and its impact on wildlife. Participants were asked about their willingness to give an annual donation to save 50,000 offshore Pacific seabirds from the consequences of small oil spills. One group was given no anchoring question, but the other two groups were asked "Would you be willing to pay. . . ?" before being asked "How much would you be willing to contribute?" The responses are summarized in Table 7.1.

Table 7.1 Responses to the Anchoring Demonstration

Anchor	Average Response
No Anchor	$64
$5	$20
$400	$143

Source: Adapted from Kahneman, 2011, p. 124.

Anchoring can become more entrenched if other biases are present, for example:

- Confirmation bias—Whereby the estimating team seek data which reinforces the initial estimate.
- Availability bias—Whereby the information readily available is sourced and assumed to be of high value for the task at hand. This leads to limited in-depth investigation and analysis.

For example, a senior manager declares that he believes the project should be completed within 18 months at a cost of $1 million and will realize $2.5 million in benefits. Information is sought to corroborate these figures, and even if the evidence suggests that these figures are incorrect, it will be difficult and perhaps career-limiting to suggest that the senior manager was wrong. So, information is found which supports this perspective, and confirmation and availability biases have played their part in helping to make a suboptimal decision.

Rule Beating and Unintended Consequences

As discussed earlier, in Chapter 1, selecting a metric or offering an incentive can often result in unintended behaviors and actions. Meadows (2008) refers to these circumstances and workarounds as "rule beaters," where users and stakeholders apply the letter of the law or instructions, but perhaps not the spirit, in a way which is advantageous to them. A couple of examples are provided, which may be familiar.

- Government authorities spend money toward the end of the fiscal year to avoid losing funding (equivalent to their underspending of this year's budget) in the subsequent year. This creates the behavior of spending for the sake of using the available funds, or allocating the money as if it was spent, and not necessarily providing value while ensuring that the next year's budget is maintained in full.
- In the 1970s, the U.S. state of Vermont enacted a land-use law (Act 250) which required a complex and lengthy approval process to subdivide land into lots of 10 acres or less. The result is that there are now a large number of lots which are just over 10 acres in size. In order to subdivide land and avoid the approval process, landowners split their lots into parcel sizes which were just above the limit for applying Act 250.
- The U.S. Endangered Species Act restricts development wherever an endangered species has its habitat. Some landowners, upon discovering that their property is home to an endangered species, will hunt or poison it to allow the development of the land.

There are numerous examples of similar behaviors, where good intentions are wasted because the resultant behaviors of the stakeholders lead to different results than those intended. Some of the examples are quite humorous, when viewed by an outsider; perhaps the most famous and most comical example comes from Horst Siebert (2002), who popularized the term "the cobra effect," which can be summarized thus:

> A British governor in colonial India wanted to rid the city of snakes—there were too many cobras in the city. A reward was offered for every dead snake presented, which, it was intended, would solve the problem as people hunted these pests and killed them. However, some entrepreneurs saw an

opportunity and began to breed the snakes so there would be an endless supply of snakes and associated revenue stream. Suddenly, the governor and his administration were swamped with dead snakes and claims for rewards. The scheme was withdrawn, as it was realized that it was being abused. Then, there was another problem—a number of cobra breeders had a population of now-worthless snakes. The breeders had no desire to keep the animals and feed them when there was no market for their sale, so the snakes were released from captivity. This created a larger wild population of snakes than there was initially and caused more consternation.

There are other examples of similar schemes, which resulted in unintended results that made the original problem worse. Vann (2003) documented the tale of rats in Hanoi, which followed the cobra effect almost identically. The Hanoi story had one twist: The reward was paid for the tails of the vermin. Enterprising rat catchers would catch the rats, cut off the tails, and release the rats so they could reproduce, thus sustaining the population. It appears that there will always be a group of people who see an opportunity to behave in an opportunistic manner or unintended way to bend or beat the rules to the detriment of the sponsors.

This behavioral issue is perhaps among the most important of the biases, because it affects the outcomes and benefits in a significant way. The result may be a misunderstanding of the changes or an intentional misinterpretation in order to take advantage of the opportunity. In either case, the result may come as a surprise to the team and the sponsor. With the long-term nature of programs and changes, it is essential that the team gather and process data and observations regarding the behavioral changes to allow rapid response to any anomalies or deviations from the plan; the importance of the monitoring and feedback mechanisms cannot be overemphasized.

7.1.2 Combatting the Biases

Reference Class Forecasting

Project managers and estimators are familiar and comfortable with the term "analogous estimating." This is the technique of applying the estimates and actual performance metrics from one project to the planning activities of a new but similar project. Of course, it requires setting aside the uniqueness of each project and accepting that there are features of commonality in the work required. One of the issues with using analogous data is the lack of availability of reliable information. Kahneman (2011) used the term "outside view," meaning an external perspective which is devoid of the assumptions, prejudices, and biases that are common within project teams and organizations.

Reference class forecasting is the technique used to apply that "outside view," through the development of a data set which is representative of the type of work being undertaken. Assumptions are replaced by data gathered from the real-world examples of a group of similar projects. The *reference class* is the group of similar projects or activities to which the data can be applied. Obviously, there is cost and effort associated with gathering, analyzing, and presenting the data. So, the selection of the reference class should be carefully considered to ensure that it will be possible to gather a reasonable, representative, and reliable population of data, and that the data will be applicable in enough future projects to be useful.

As discussed earlier, human judgment tends to be optimistic, and there is a need to moderate estimates and forecasts with a healthy dose of reality. Kahneman and Tversky (1973) concluded that single-point estimates with no regard for a distribution of potential results was the major source of forecasting error. In scheduling, this has been recognized (to an extent) by the use of three-point estimates and statistical techniques such as Monte Carlo simulations. Reference class forecasting moves beyond this by assembling a database of proven values from similar projects as the mechanism for demonstrating the accuracy of estimates.

Kahneman (2011), whose work led to the award of the Nobel Prize in Economics in 2002[*], stated that reference class forecasting was "the single most important piece of advice regarding how to increase accuracy in forecasting" (p. 251).

Reference class forecasting comprises three steps:

1. Identify similar projects, for which data are available, to form the reference class.
2. Establish a probability distribution for the reference class for the attribute for which the forecast will be required.
3. Compare the project with the reference class to calculate the most likely outcome.

The use of a reference class is more than looking at the projects within the organization. It requires the collation of information and data from projects

---

[*] Amos Tversky, the long-time collaborator of Kahneman, unfortunately died in 1996, making him ineligible to share that Nobel Prize.

conducted elsewhere, so that the complete range of possible results can be assessed. As a result, it takes time and effort to establish a database which clearly demonstrates the probabilistic range of results that can reasonably be expected. The team can then analyze these data and determine the range of likely outcomes in terms of both delivery (cost and schedule) and benefits. Although projects are defined as unique undertakings, it is possible to find a reasonable and reliable reference class by examining the major activities and phases of projects and programs.

Wisdom of Crowds

Project managers are encouraged to be inclusive when planning and especially when estimating for those plans. There are a number of reasons why this is the case. First, there is a need to refer to subject-matter experts when compiling an estimate, and the broader the experience of the collective group, the more likely we are to consider all of the possible results and obtain information covering all eventualities. Second, there will be risks, which people have been exposed to in different circumstances, that will need to be factored into the estimates. Finally, the inclusive nature of estimating and planning helps to gain commitment to the goals set within the plans which are generated.

A number of techniques can be used for gathering the information from the groups. One of these is the Delphi technique, in which participants are asked to make their contributions to a workshop or process without reference to the other participants. The results are compiled and analyzed to establish the predicted range of results or estimates, which arise from the collective experiences of the group but without one person influencing the others. One disadvantage of this approach is its reliance on the memories of the personnel involved, which could be clouded by some of the biases discussed above. Another disadvantage is that for the Delphi technique to have greater value, several rounds of data collection tend to be needed to derive a common solution, which all the participants agree is the best one available.

Surowiecki (2005) referred to the "wisdom of crowds" when discussing these methods. It has been demonstrated that the estimates of a group are more reliable than individual estimates. Individuals may provide extreme estimates, outliers, which as a single point of reference may be unrepresentative, whereas the group tends to provide the context around each result. The "trick" is finding the correct group—the group should not be homogenous but needs to bring a breadth of experiences, which can add value to the collective thinking. Homogenous groups are likely to tend toward groupthink and provide some false sense of security and comfort that the activity has been undertaken, and the result may be relied on completely.

7.1.3 Calculating the Value of Benefits

Three primary methods for quantifying the worth of the benefits when developing a business case can be used to make decisions regarding the initial and ongoing investment:

1. Net present value (NPV)
2. Payback period
3. Benefit–cost ratio (BCR)

Net Present Value (NPV)

Net present value takes into account two important aspects of the investment:

1. The timing of payments and incoming revenue—Future payments are discounted to their present value, that is, as if they occurred at the start of the investment. This provides a fair comparison between the outgoings and the income (benefits), taking into account the time difference between the two events.
2. The initial investment and the ongoing costs are considered so that the benefits are an accurate reflection of the changes.

Present Value (PV)

The value of money changes over time; it is devalued by inflationary factors. This means that $1 million in 2018 is valued more highly than $1 million in 2025. When considering long-term investments, this can have a significant effect on the business case. The present value (PV) of future payments or income is calculated to convert the value of all future transactions to a common time frame, that is, to create equivalent values regardless of when the transactions occur and thus make them directly comparable. The mathematical equation for calculation the present value is

$$PV = FV/(1+i)^n$$

where

i represents the discount rate, which is the rate of return that could be earned on an investment in the financial markets with similar risk—it is the opportunity cost of the funds.

130 Implementing Project and Program Benefit Management

    n represents the number of periods (months/quarters/years) over which the PV is calculated. Generally, the calculation is performed for fiscal years.
    FV is the future value of the amount.

So, for the example above, consider the present value (in 2018) of $1,000,000 in 2025:

    n = discount rate, assumed to be 5%
    n = number of periods, 7 years
    FV = future value to be reduced to its present value, $1 million

Then

    PV = 1,000,000/(1 + 0.05)^7
       = 1,000,000/1.4071
       = 710,681

In other words, $710,681 in 2018 is equivalent to $1,000,000 in 2025.

The calculation of PV is important for comparing one value with its current equivalent. Net present value extends this to the more complicated situation where costs and income (savings) are spread over several periods of time.

The NPV represents the total of a series of cash flows (incoming and outgoing) of the present values (PV) over the investment period and the period when the benefits are realized. It compares the value of the investment at today's value with the net income in the future, adjusted to account for inflation and returns.

Calculating the NPV

The formula for calculating the net present value may be written as

$$NPV(i,N) = \sum_{t=0}^{N} R_t / (1+t)^t$$

where

    i represnts the discount rate, which is the rate of return that could be earned on an investment in the financial markets with similar risk—it is the opportunity cost of the funds.
    N represents the number of periods (months/quarters/years) over which the NPV is calculated. Generally, the calculation is performed for fiscal years.

t the time of the measurement of the cash flow.

R_t is the net cash flow during the period, equal to cash inflow – cash outflow.

R_0 is often used to refer to the net cash flow during "Period 0," that is, t_0, which is the period when the initial investment is made (and is therefore a negative value)

Applying the formula to an investment situation, where

Initial investment	$1,000,000	R_0
Year 1 net cash flow	$300,000	R_1
Year 2 net cash flow	$400,000	R_2
Year 3 net cash flow	$500,000	R_3
Discount rate	5%	i

Then

NPV = –1,000,000 + [300,000/(1 + 0.05)^1] + [400,000/(1 + 0.05)^2] + [500,000/(1 + 0.05)^3]

NPV = –1,000,000 + 285,714 + 362,812 + 431,919

NPV = $80,445

The sum of costs and net cash flow (without discounting the values of years 1, 2, and 3 revenues) indicates a return of $200,000, which may create a false sense of viability for the initiative. The NPV moderates that value, providing a better comparison which takes into account the lag between investment and return, which in this case is about 40% of the simple sum of the incomes. The discount rate used in this instance is quite modest, but it clearly demonstrates the effect that time can have on the viability of an investment.

The Discount Rate

The selection of the value of the discount rate is entirely a matter for the organization's executives and sponsors committing to the initiative. From the above, it can be deduced that the higher the discount rate, the greater the benefits need to be in future years in order for the investment to remain viable. The higher the discount rate, the greater will be the erosion of value in future years.

To avoid arbitrary figures being applied, the discount rate can be based on a forecast of the inflationary rates, the Consumer Price Index (CPI), or similar predictive metrics. Alternatively, consider a discount rate which is equal to the return that could be achieved by investing the funds in an alternative venture which carries a similar risk. This allows the comparison of different projects and also the "Do nothing" approach, which is the option not to commission the project but to invest the funds in other opportunities or markets.

When considering large projects with long-term outlooks with respect to realizing benefits, choosing a single discount rate and expecting that to remain constant for the complete life cycle may be optimistic and provide a misleading result. It is acceptable to change the value of the discount rate (i) for different years to accommodate the best estimates of future conditions. This can be applied to build a more realistic view of the future, taking into account economic projections and levels of certainty and risk. Obviously, this changes the formula above, and the calculation becomes more complicated.

Making Decisions Based on the NPV

The NPV is an indicator of the increase, or decrease, in value an investment will provide to an organization. It is important to recognize that the NPV may be negative, which will indicate that the sponsor will be worse off at the end of the realization-of-benefits period. This may occur if the initial investment is enormous; the benefits are spread over a lengthy period and are therefore eroded by their discounted value. An example of such an investment might be a nuclear power station, where the initial investment is massive and the revenue stream providing the return is distributed over a very long term. Depending on the discount rate selected (the returns available from other sources), a better return might be found elsewhere. But power-generation corporations build power

Table 7.2 Summary of the Broad Interpretation of NPV

If...	Which means...	And...
NPV > 0	The investment will add value to the sponsor.	The project may be commissioned.
NPV = 0	The investment will add no value to the sponsor, nor will it reduce value.	An objective view of the project will be neutral regarding this investment. However, it should be noted that a small change to the plan could result in a positive NPV, and a small deviation from the plan could result in a negative NPV being realized.
NPV < 0	The investment will subtract value from the sponsor.	The project should be rejected because it will detract from the organization's value.

stations—so often this defines the type of investment, which is supported by other influencing factors and not based merely on the NPV.

When comparing the viability of individual projects and identifying the "best" project to undertake, some relatively simple guidance may be applied. A project is viable, that is, it will return value to the sponsor, if the NPV is positive. All other considerations being equal, the "best" project is the one with the greatest value for NPV. This assumes that the funding for all projects is available, and the projects all have similar levels of risk. As an extension of this conclusion, the best combination of projects is the one which will have the greatest combined NPV and can be delivered within the funding limitations. Table 7.2 summarizes the simplistic interpretation of NPV for investments.

Payback Period

Payback period is a simple method for measuring the overall value of an investment. The *payback period* is the time required to recover the costs of the investment, thus reaching a breakeven or zero-sum point. The payback period is usually expressed in months or years and can be used to compare investments of similar magnitude. When comparing two similar investments (strategic significance and other comparisons being equal), the one with the shorter payback period offers the better result.

In a simple situation, where the investment leads directly to a cost saving, or an increase in revenue, the payback period is calculated by determining when the cost savings (or increases in revenue) equal the initial investment, that is,

Initial investment = savings (year 1) + savings (year 2) + savings (year 3) + . . .

As an example, an organization invests in new software costing $1 million. The implementation of the new software saves $250,000 per annum.

Investment = 1,000,000
   = 250,000 + 250,000 + 250,000 + 250,000
From year   1   2   3   4

In this example, the payback period is four years.

However, life is rarely so straightforward, and there are likely to be ramp-up effects to consider, and the depreciation of the value of money over a period of four years may be large enough to influence the decision making.

In more complex situations the payback period is calculated by determining when the initial investment equals the cumulative net cash flow. In this instance the net cash flow is considered as

Initial Investment		Year 1	Year 2	Year 3	Year 4	Year 5
$1,000,000	=	250,000 – 50,000	250,000 – 50,000	250,000 – 50,000	250,000 – 50,000	250,000 – 50,000
$1,000,000	=	200,000	200,000	200,000	200,000	200,000

Figure 7.3 Example of the Calculation for Payback

Net cash flow (for a period) = cash inflow − cash outflow

Using the example above, with $1 million investment, and $250,000 in cash savings each year, there may be ongoing licensing costs associated with the software, for example, $50,000 per annum, as shown in Figure 7.3.

In this example, the payback period is five years because the annual licensing costs of the software reduce the net savings by 20%. It is tempting, but flawed, to claim all of the savings and ignore any "operational" costs because the business case will be stronger, and the payback period will be shorter, making the investment appear to be more attractive.

When comparing the two examples above, if an organization can afford only one initiative at $1 million, then the first example is the more attractive. This assumes that all other factors, including strategic alignment, urgency, risks, and so on, are equal. The first example is more attractive because the payback period is shorter. It takes the investors four years, instead of five, to recover their investment.

When to Use the Payback Period

The payback period is a relatively simple analysis, which has advantages and disadvantages. The technique does not factor in the changing value of money over time, that is, it assumes that a dollar spent today during the investment period is equal in value to a dollar saved in five years' time. Given the economic climate and inflation, this tends not to be the case, and the value of money depreciates over time. So, the payback period is not a good tool to use for long-term investments or in situations where inflation is high, and where alternative investment opportunities offer a significantly higher return.

However, this technique may be a useful tool for making comparisons between projects which have a relatively short payback period, for example, one that may be measured in months rather than years. For example, a project may be completed within one financial year with the benefits realized immediately and measurable in the current or following financial year. The benefits may accrue monthly. In this instance the period of time between making the investment and realizing the benefits is small, and the calculation of the payback period is robust.

The payback period may be a useful tool when:

- The gap between investment and benefit realization is relatively short, making the time effect of the value of money less significant.
- The project/investment is in the early stages of development, and a quick comparison is required between two or more investments.

Benefit–Cost Ratio (BCR)

The benefit–cost ratio (BCR) is an indicator of the overall value for money of the project, and it is calculated by dividing the total value of the benefits realized from the changes created by the project by the total cost of delivery and transition. Similar to the calculation of the NPV, the benefits value used in calculating the BCR should be expressed in discounted present values.

This approach is popular because the sponsor and other key stakeholders can state with some assurance, for example, that for every dollar invested, $1.50 of benefits will be created. This is a simple metric which provides great reassurance to investors or warns them of potential hazards. It is most accurate when all of the benefits can be expressed in financial terms. If that is not possible, this approach will downplay the benefits, and the BCR is likely to be close to or below 1.0.

As with the NPV, a BCR of greater than 1.0 indicates a value-adding investment, and the greater the value of the BCR, the better the investment. Table 7.3 summarizes the interpretation of the BCR.

Table 7.3 Summary of the Broad Interpretation of BCR

If . . .	Which means . . .	And . . .
BCR > 1	The investment will add value to the sponsor.	The project may be commissioned.
BCR = 1	The investment will add no value to the sponsor, nor will it reduce value.	An objective view of the project will be neutral regarding this investment. However, it should be noted that a small change to the plan could result in a positive BCR, and a small deviation from the plan could result in a negative BCR being realized.
BCR < 1	The investment will subtract value from the sponsor.	The project should be rejected because it will detract from the organization's value.

7.2 Assessing

These financial techniques are useful tools in demonstrating the value of the investment. However, they can only cover those benefits which can be quantified in monetary terms. Complexity increases when there is a mixture of financial and nonfinancial benefits to be gained. Ultimately, someone needs to make a decision regarding the investment: Is it viable?

When the investment of $10 million results in benefits worth $15 million being generated, the decision is relatively straightforward. Of course, the duration of both project delivery and benefit realization must be taken into account, as well as risks and the volume of change, but a BCR of 1.5 is encouraging.

Similarly, an investment of $10 million which results in benefits worth $5 million being generated, meaning that the BCR is 0.5, is likely to result in a quick and simple decision.

However, if the investment of $10 million results in financial benefits worth $7.5 million being generated, plus some nonfinancial benefits, the decision is compounded by the mixture of benefits. The business case is no longer a simple comparison of apples with apples; the stakeholders may be asked to compare apples with some apples and some oranges and a few bananas thrown in for good measure. The decision is complicated by the mixture of financial and nonfinancial gains. The sponsor needs to make a decision about the perceived value of those benefits, and, in this case, whether they exceed the $2.5 million required for the BCR to equal or exceed 1.0.

Remember that a BCR of 1.0 means that the financial benefits are equal to the costs. Ultimately, someone within the organization must make a decision as to whether the nonfinancial benefits are worth the investment.

It is correct to say that benefits do not *need* to be financial. There are many examples of nonfinancial benefits which can legitimately be claimed, such as

- Fewer steps in a process
- Reduced emissions of noxious or dangerous chemicals
- Increased hits on a website

However, if these can be translated into legitimate financial values, it will make the assessment of the investment much simpler. The stakeholders will be back in the position of comparing apples with apples.

In terms of practical advice for the practitioner/project manager, if it is possible, the benefits should be stated in financial terms. This simplifies the argument regarding whether to invest or not. This may mean that the nonfinancial benefits need to be redefined or clarified. Often it means forecasting over a longer timeframe.

Examining one of the nonfinancial benefits above, assume a government agency undertook a transformational program to encourage clients (being the public at large) to undertake transactions online, rather than by going to an office in some centralized location, and one of the key objectives was to simplify the process of engaging with various groups within the agency. By simplifying the process, there would be fewer steps in each process to undertake the transaction. This was clearly an important improvement for some of the stakeholders and measurable and therefore a nonfinancial benefit. However, looking into the longer term, this is not the end of the benefit: There is a pathway linking this initial benefit with several other benefits and changes in behavior.

Figure 7.4 shows the complex nature of benefits management and the dependencies among benefits. By way of explanation, starting with having a simpler process:

- Fewer steps are involved in making the transaction (Benefit and also an Outcome).
- If there are fewer steps in the process, more people are likely to make the transaction online. More people using the online payment approach is a measurable benefit. There will be an immediate benefit as early adopters take to the new method, and the numbers of users will grow over time hopefully (Benefit).
- Several of these results will not happen automatically. There must be an education process, and a marketing and advertising campaign to advise the audience of the new service and how to take advantage of it. These changes within the organization are essential to the overall success of the project, and they must be funded. It is important that these activities are identified and costed in the business case to ensure that the right and adequate resources are available at the end of the project or following completion of the project. These activities may extend over a considerable period of time to ensure that the target audience knows about the new option and changes its behavior from the old way of working to a new "business as usual."
- Fewer people go to the office to undertake the transaction. This could be a major benefit to several stakeholders and may result in further cost savings in the longer term. If more transactions occur online, there will be fewer people in the office waiting to be served. This will allow the pattern of work to change, resulting in costs savings (Benefits) through reduction in office staff required to serve clients. This benefit could be realized as a cost saving through fewer staff or by maintaining the size of the staff but redeploying them to other productive work. This benefit can only be realized once a threshold of users has committed to the new system, and there needs to be a critical mass of users.
- Fewer face-to-face transactions may also result in the option to reduce office space and save rental expenses (Benefits) or reduce the number of offices required to serve the community.
- Closing offices may cause a dis-benefit to be realized—for some of the stakeholders in the community, if some offices are closed, going to an office may take longer and cost more, and it will *encourage* them to perform the transaction online.

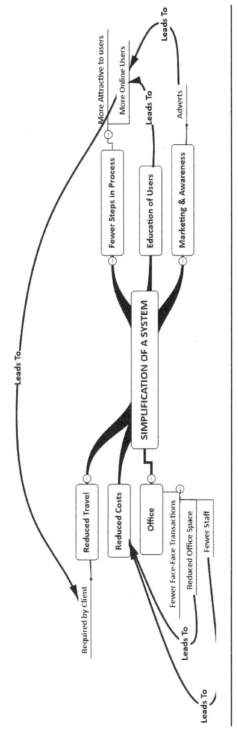

Figure 7.4 Mindmap Showing the Connections between the Benefits

- Fewer people travel to the office. Reduction in traffic (Benefit, but may be difficult to measure and Attribute to this project, especially if other infrastructure projects are being conducted at the same time) is one benefit which may be omitted from a business case, due to the difficulty of attributing the benefit to the changes created through the project. If the current office is in the middle of the city, the central business district, there may be little observable difference in traffic flow. If the current office is outside the city, it may be obvious that traffic patterns have changed. Perhaps easier to calculate would be the time of travel saved; hopefully there would be a reduction in the number of face-to-face transactions performed, and this could be compared with the baseline performance level. There may be some data which exist or can be gathered to establish the baseline and follow the change that captures the travel time for clients.

At some point, decisions need to be made regarding the following:

- Which of these benefits will be measured and quoted in the Business Case document? Some of the benefits will be substantially greater than others; in fact, some will not be worth measuring because of their size.
- When will benefits be measured?
- When will the measurement of benefits cease?
- At some point the benefits will be influenced by other, possibly external, events, and it will no longer be reasonable to attribute them to the project.

Some benefits will be realized well into the future, when the project is a distant memory. The benefits may continue but are embedded into the method of work/business as usual, so their relevance as an improvement from a previous point is no longer valid:

- Benefits may be realized in the first few years following the completion of the project, but then wane and reduce over a period of time such that they become less significant and not worth the effort of monitoring and measuring.
- During the review and assessment of the forecast benefits, the project sponsor, and key stakeholders, should determine the parameters and constraints of measurement regime. These limits will be recorded in the benefits realization plan.

Quantifying all of the benefits in financial terms is not always desirable because of the context of the investment. The message may be corrupted if the single focus of a project is the financial return. For example, a project aimed at improving safety in a workplace should not focus on the cost savings as the primary benefit, although there may be some significant cost savings to be gained (Duff et al., 1994). In reviewing labor productivity on construction sites in the UK, they reported that sites which were deemed to be "safe" had significantly higher labor productivity. Labor expenses account for approximately 30% of total construction costs, and an improvement in productivity will reduce overall costs and (for the contractor) increase profitability. However, to commit to a safety improvement program with the intent to reduce costs sounds callous and unfeeling. For such a program, having metrics such as reduced number of lost-time incidents or increased time between incidents would be more relevant and appropriate. It is better if these nonfinancial benefits can be linked in some way to an organizational goal or priority, so that the investment can be argued in the context of the corporate values.

7.3 Documentation

7.3.1 Benefit Profile

A benefit profile is a document which contains the information about a single benefit. Each recognized benefit, generally meaning those which contribute toward the business case, has its own profile, which should be amended and updated throughout the project. Following the assessment process, the benefit profiles need to be updated and supplemented with the information generated during the calculation of the value of the benefits.

7.3.2 Benefits Realization Strategy

The purpose of the benefits realization strategy is to ensure that there is a single point of reference and common understanding regarding the "ground rules" for the management of benefits within the project and following its completion. During the process of assessing the benefits, the benefits realization strategy should be reviewed and referenced. It may even be amended if, for example, priorities change.

7.3.3 Business Case (Initial)

The business case is the primary decision-making document for the key stakeholders. It presents the case for the investment and its continuation as the project progresses. Therefore, it must be a live document, which is maintained using the best available information from the project and its environment. The document itself has a relatively straightforward template which covers:

- The cost of delivering the project and the subsequent changes required to achieve the benefits.
- Timeframe with significant milestones and constraints recorded.
- Benefits including their forecast magnitude and the timing of their realization.
- Risks—Generally only the major risks are listed in the business case document, with a summary of the total magnitude of risk.
- Comparison of the costs and benefits taking into account the risks involved.

7.3.4 Benefits Register

The benefits register is a summary of the benefit profiles. It is a list of the benefits with details of the attributes of each one. At any point at which the benefits are identified or reviewed, the benefits register should be updated.

7.4 Reviewing and Decisions

The completion of the process of assessing the benefits should coincide with the completion of the initiative's planning phase. This allows compilation of the documentation and information which is needed from the project in relation to benefits realization, to make the investment decision whether to continue with the initiative. The decision to invest in the change initiative should be based on:

- The initial business case document
- Assurance conducted within the organization to review and make recommendations regarding the project
- Independent reviews/gateways conducted to bring in external expertise, which is not biased in its perspective.

7.4.1 Initial Business Case Document

One of the most important things to remember about the business case document and the concept is that it is live. Things change. As the project progresses, costs, timeframe, and the allocation and availability of resources will alter.

Pessimistically, this translates to the project being later than expected, delivered at a cost greater than the budget, and/or resources not being available, which could affect the cost and schedule.

In addition, the forecast of the expected benefits may change over time. Internal and external factors may influence that forecast. Taking a hypothetical example, a UK-based manufacturer may have invested in new facilities in 2014 based on forecasted exports to other countries within the European Union. Export sales may have worked out well initially, say, in 2015. Sales projections (benefits) may have been adjusted during 2015 as the possibility of a vote to leave the EU became more visible and real. In 2016, those sales figures would likely have been adjusted downward following the "BREXIT" result. In 2018, those forecasts might have been adjusted upward as the impact of the withdrawal from the trading partners was understood, and the negotiated terms of the split became known (and were, perhaps, less onerous than feared initially).

In short, as the circumstances within the project become known more accurately, is a need to adjust the estimates within the plans, and the associated documents. Similarly, as the environment outside the project and its organization change, there is a need to reevaluate the plans, and especially, the business case document. When the decision is made to close programs and projects, it is based on the business case, with the realization of benefits being the key focus.

The information available at the end of the process of assessing the benefits provides many of the inputs for the business case document at this point. The document is based on the best information available at that time it is created. However, this document, in its initial state, is prepared using early estimates and may not be entirely accurate because:

- The estimates may be rushed, to get the project started.
- The estimates are often prepared while, or before, the scope of the project is developed.
- The team and those involved in assessing the benefits may not yet be familiar with the project, its objectives, and the environment. This understanding will develop as the project progresses.

The term *initial business case* is often used to denote the point of the project life cycle at which the document is prepared.

7.4.2 Assurance

Assurance is broadly applied to the review of the project, its documentation, and information by people and teams who are not directly involved in the work. The primary purpose of assurance is to provide comfort to the decision makers that the information they are relying on is correct. Assurance is often viewed as a part-time and intermittent role, which is called on as required. In practice, assurance is not always independent enough of the project to make an objective assessment and provide constructive advice. General approaches to the provision of assurance include:

- Engaging a project management office (PMO) or similar corporate group to provide the assurance.
- Use of an existing body is a valid approach to assurance. The primary issue is that the group conducting the assurance process may be too close to the work being undertaken and is not able to render truly independent advice. The PMO may have succumbed to the same biases and heuristics as the delivery team and will not necessarily be able to separate the precise status of the project from the assumptions made.
- Conducting independent reviews. Independent reviews may be conducted at any time during the project and beyond. The use of a fresh set of eyes to offer an unbiased opinion and review of the initiative may be useful on a daily basis.
- Creating and implementing formal review points or gateways.

7.4.3 Independent Review

There are points in any development life cycle at which it is prudent to undertake an independent review of progress to date and plans for the future. The end of the process of assessing the benefits coincides with the completion of an initial business case document and its review and evaluation. A decision is made regarding the validity and viability of the business case, culminating in the project halting or progressing. This is a major decision. In fact, halting projects at this point could be extremely valuable as investments are directed toward viable projects where the stakeholders have a clear understanding of the objectives and the return on their investment. For projects which are authorized to proceed beyond this point, this represents a milestone, or gate, which requires a reasonable input of resources to move past. In either case, casting a fresh set of eyes over the project to confirm the accuracy of the estimates before making this decision is wise.

Considering the potential scale of the investment engaging external consultants or internal reviewers to assess the plans should be considered a wise move. The costs associated with such a review should be factored into the planning of the work and will likely pale into insignificance in the overall project. Such a review will provide confidence to the stakeholders and the project team and increase the likelihood of the right decision being made.

7.4.4 Learning Lessons

It is important that lessons from previous experiences and initiatives are heeded and applied as early as possible. Failure to apply these valuable insights may lead to early, directional decisions being made which are difficult, or costly, to recover or roll back.

The application of lessons can, and should, be applied throughout the initiative's life, and the benefits life cycle, but is particularly effective when applied early during the planning phases. It is interesting to note that seeking and using lessons features as the second activity within PRINCE2® (AXELOS, 2017)—the first activity being the appointment of the Executive and the Project Manager—and as an input into all planning-related activities within the *PMBOK® Guide* (PMI, 2017a).

Lessons are available within the organization and the team. However, there are often lessons, which can be learned from others. Utilizing other projects' successes and failures is an inexpensive and easy way of learning.

Reich (2004) found five "knowledge traps" which apply to IT projects:

1. Lessons learned—"Without access to lessons learned from comparable projects, the team will have lost an important opportunity for a quick and informed start."
2. Team selection—It is important to have ". . . all relevant knowledge areas included in the team or available. . . . Problems arise when the project manager cannot select his/her team or when the knowledge profiles of team members are not available during the selection process."
3. Timing the entry and exit of team members—". . . members added to a team after the mid-point of a task rarely change the team's direction, so care has to be taken to have key knowledge inputs early."
4. Volatility in the governance team—There is a knowledge-building process within the governance team and, through that, with the key stakeholders. Volatility among this group may influence project resources and decision making.
5. Lack of knowledge among the governance team—Senior executives become involved in initiatives because the outcome is important to their role. Without experience, training, and support, the sponsoring group may not be familiar with their role and when they can have greatest impact on guiding the program manager and business change managers.

To combat these traps, Reich advises that the organization should:

- Establish a "knowledge sharing center."
- Establish and maintain knowledge levels.

Table 7.4 Activities Conducted within the "Assess the Benefits" Process

Activity	Purpose	Responsibility	Documentation
Workshops with the program team, the business change managers, and subject-matter experts	• Quantify the benefits: • Establish the scale and significance of each benefit.	Program manager (coordinating the team)	Updated benefit profiles
Workshops and other data-gathering techniques/research	• Identify lessons which can be applied during this process, and later in the life cycle.	Program manager	
Review of benefits	• Confirm the application of lessons and avoidance of bias.	Assurance	
Compile the initial business case document	• Compile the initial business case document.	Program manager	Business case document
Review the initial business case document	• Confirm that the assumptions, logic, and calculations in the business case document are reasonable and accurate. • Confirm that the business case is viable.	Assurance and sponsor	
Approve the initial business case document	• Confirm that the business case is acceptable to the key stakeholders.	Sponsor and sponsoring group	

- Create channels for knowledge flow.
- Develop "team memory."

The team will include a wealth of experiences, and these should be drawn on as a starting point. Workshops and interviews are helpful to uncover useful lessons from the team. It may be useful to hold a series of short workshops involving different levels of the team to ensure that each group has an opportunity to state their thoughts without pressure from superiors. Additionally, there will be different perspectives and therefore different lessons for each level of the team; for example, the sponsoring group may have lessons gleaned from their experiences, which place greater emphasis on the changes within the organization and the benefits. The sponsoring group members are less likely to be interested in lessons which relate to technical matters.

Flyvbjerg and Budzier (2015) promote an approach of simplification of the initiatives, rather than attempting to fully model the complexities. They advocate decomposition into smaller and more well-defined units of work because "leadership will not succeed by modelling complexity; they will succeed by understanding simplicity" (p. 22). In line with that, Table 7.4 decomposes the process, "Assess the Benefits," into its activities.

7.5 Summary

Assessing the benefits involves gaining an understanding of the magnitude and timing of those benefits which have been identified. Each of the benefits must be quantified, preferably in financial terms. If placing a financial value on a benefit is not possible, or desirable, the scale of the benefit must be established so that it may be incorporated into the business case document and the associated discussions and decisions.

The use of net present value as a determinant for making decisions provides a standardized and common method for objectively comparing the investment with its returns, and one investment with others. It should be noted, however, that placing a numerical value on the NPV or BCR tells only part of the story, and other considerations may influence the decision-making processes and particularly the decision to commission a project, including:

- Funds available for investment.
- Strategic priorities.
- Relationship between one project and others—It may be necessary to undertake a project with a comparatively low BCR first, because it enables future projects with stronger business cases.

The next chapter will discuss the planning required within an initiative for the realization of benefits and their sustainability.

Some of the issues addressed at this stage should be:

- Have all of the benefits been reviewed and quantified?
- Have all of the benefits been quantified in financial terms?
- Has evidence been found to support the estimates and forecasts?
- Has an independent review been undertaken of the benefits profiles?

Exercises and Activities

Consider the data in Table 7.5.

Table 7.5

Project	Cost	Benefit				
		Year 1	Year 2	Year 3	Year 4	Year 5
A	$1,000,000	$100,000	$150,000	$200,000	$300,000	$400,000
B	$500,000	$100,000	$200,000	$300,000		
C	$1,200,000	$300,000	$300,000	$300,000	$300,000	$300,000
D	$300,000	$180,000	$120,000	$60,000		

Assume that the risk profile of each project is similar and that they are all closely linked with organizational objectives.

1. Using a discount rate of 5%, calculate the following for each project and the program, as a whole):
   - BCR (not discounted)
   - NPV
   - BCR (discounted)
2. Which project is the most attractive? Explain your reasoning.
3. Which project is the least attractive? Explain your reasoning.
4. If the budget was limited to $1,500,000 overall, which project(s) would you recommend undertaking?
5. Calculate the payback periods for each project and the program using both the nondiscounted and discounted values of the benefits.
6. If the discount rate was 8%, how would that affect the investment appraisal?
7. If the economic environment was turbulent and the discount rate likely to vary over the period of the initiative, how would you advise the sponsoring group?
8. Discuss how each of the methods to calculate the value of benefits and the investment would suit your organization. Which method would be most effective? Justify your suggestions.

Chapter 8

Plan for Benefits Realization

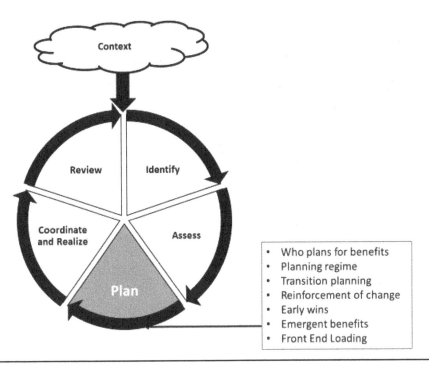

Figure 8.1   Plan for Benefits Realization within the Benefits Life Cycle

> *"No battle plan survives contact with the enemy."*
> – Helmuth von Moltke

> *"Plans are of little importance, but planning is essential."*
> – Winston Churchill

> *"In preparing for battle, I have always found that plans are useless but planning is indispensable."*
> – Dwight Eisenhower

Helmuth von Moltke meant that at the point of meeting the enemy, the plan will need to change because new events, risks, and changes will be identified which affect the original plan, and we increase the chance of success and become wiser as a result. It is the same with programs and projects, where this is referred to as *progressive elaboration*—as more information is gathered regarding the environment and situation, the better able the team is to adjust the plan, add detail to the plan, and then progress the work. It is likely that the assumptions and decisions made initially will need to be reviewed, revised, modified, and changed—and so the plan needs to be adapted, taking the new information and circumstances into account. This is perhaps even more applicable to the management of benefits, which have a longer life cycle, than the projects themselves. Figure 8.1 shows the planning process within the benefits life cycle. Planning should never be viewed as a one-time activity; it must be revisited based on information which becomes available once the plan is implemented.

Churchill and Eisenhower agreed that the plan, as a document, has limited value because it quickly becomes superseded by the situation and will be subjected to change as new information is gathered. However, the process of planning has great value because it allows the team to make decisions and think through the process of delivery. Planning is also an effective way to identify problems and risks and establish responses to them.

This chapter will explore and detail the planning activities required to make benefits realization management effective.

The *Collins English Dictionary* definition of planning is "the process of deciding in detail how to do something before you actually start to do it." Planning is vital to the success of a project or program and should be revisited throughout any initiative. Plans evolve over time as more, and more accurate and relevant, information becomes available. This leads to plans having different names or version numbers to indicate the point of development within the program life cycle or the person who owns that plan. The volume and accuracy of information available to the planners relates directly to the point in the program life cycle. In practice, this means that plans should be revised when new and updated information becomes available.

For me, planning is an opportunity—an opportunity to undertake a dry run of the project while incurring relatively little cost and with minimal risk. Planning allows the team to develop a roadmap for the intended workload, showing how the project will be undertaken. The team will learn from this process. It is much cheaper, and less embarrassing (because problems and errors will not be visible to all), to make the errors while planning rather than during the delivery of the project. The development of the plan should be an iterative process because the planning process will raise questions, which will be considered and addressed, and identify obstacles to be removed or avoided. As targets and goals are committed "to the page," risks and issues will become apparent, and they will be logged or addressed to allow for smooth passage.

It is not possible to plan for the realization of benefits in isolation; this planning must be integrated into the planning activities associated with other elements of the program. The timing of project activities has a direct impact on the planning for engagements with those stakeholders affected, the planning of activities related to changes required within the operational environment, and the timing of measurement activities required following the completion of the project. The following activities need to be planned, resourced, and scheduled:

- The work involved in the delivering the project and its contents
- The communication activities required to prepare for the change
- The work involved in creating the change required once the project is completed
- The work involved in the transition from the current state to the future environment in the community or workplace
- Reinforcement activities which are anticipated, to ensure that the changes stick and do not slide back to the old way of operating
- Activities involved in monitoring and measuring the benefits and providing governance to the process
- Activities involved in sustaining the benefits

8.1 Who Needs to Be Involved in Planning Benefits

Planning should not be a solitary task. The best results are achieved when a group of people collaborate and use their combined experiences to identify, estimate, and remove obstacles to progress and success. This has been referred to as the "wisdom of crowds" (Surowiecki, 2005), whereby a group of people is more likely, as a collective, to avoid the biases involved in estimating and decision making which are exhibited by individuals. A group of people, especially those with previous experience, can bring their aggregated wisdom to bear when estimating, resulting in more realistic data being produced. However, it is important that the group, or individuals, are encouraged—indeed, forced—to consider all of the information they have gleaned through their experience and avoid "groupthink." To avoid groupthink, it is important to consider the thoughts and opinions of a diverse group from varied experiences and technical backgrounds. There is evidence that even when people have information which could be applied to the current situation or plan, they do not always use it. This is one of the reasons a written record of previous, similar work is helpful. A reference group presents hard data, which can quickly be referenced and applied, rather than relying on memories. Ultimately, estimating and forecasting boils down to picking a number (or a range of numbers) which represents the collective and considered opinion of the estimators.

Not all of the planning will be undertaken by the program manager, and others must become involved and take responsibility for the planning required. Generally, the program manager will facilitate the process of planning and coordinate the documentation and results. It is likely that other members of the team, and possibly some subject-matter experts brought in specifically to assist with the planning activities, will have more information regarding the technical, changes, and operational environments and will be able to provide valuable input into the plan as it relates to those areas.

8.2 The Planning Regime for a Benefit Life Cycle

Effective planning practices should be applied whenever planning activities are required. Four aspects of benefits realization management will be subject to planning:

1. *Program and project planning,* which will cover the delivery of the program and its internal monitoring and control, which will be used primarily by the program manager, the project manager, and the delivery team.
2. *Benefits realization planning,* as it incorporates reviews, the measurement of outcomes and benefits, and the feedback and reporting mechanisms. The output is a plan of activities needed once the new system becomes operational, to ensure that the benefits are realized and measured.
3. *Transition planning,* which addresses the change management elements of the investment and the introduction into service of the new system, and the removal of the existing one. This will be used primarily by the business change manager (BCM) to manage the change-related activities.
4. *Sustainment planning,* which addresses the longer-term issues and activities required to sustain the new environment and regime. It may include

corporate activities such as adjustments to recruiting strategies and policies which need to be implemented to ensure that the new business as usual (BAU) is sustainable.

In practice, these needs usually result in three separate planning documents. Each of the plans will be used by different stakeholders at different times, and different stakeholders may approve each plan. Having three separate plans allows each one to focus in detail on the relevant issues for those stakeholders who will use each one, without overloading any single document with too much information. All three plans are interconnected with dependencies among them, which must be maintained and updated as the work progresses.

Table 8.1 summarizes the purpose of each of the plans. The program/project plan will be used primarily by the delivery teams, led by the program manager or project manager, to coordinate the workload associated with the project implementation. It will contain a great deal of technical information, which other stakeholders will not always need to view. However, it will also contain the connections and milestones associated with the transitional activities and the dependencies between the project work and operational change. So, some of the information in these plans will be relevant to other members of the initiative.

Benefits realization plans are used by the individual benefit owners and the BCMs to manage the longer-term activities undertaken to ensure that the outcomes do generate benefits and that these are measured and reported.

Table 8.1 Summary of the Plans

Plan	Primary User	Purpose
Program/project plan	Program manager/project manager	To capture the activities involved in delivering the program and projects.
Benefits realization plan	BCM and benefit owners	Show the activities required to realize the benefits and identify the points at which these benefits will be measured and reported.
Transition plan	BCM	Illustrate the tasks required to change the operating environment from the existing one to the preferred future state. It will be used to communicate the change with the stakeholders affected.
Sustainment plan	BCM and other stakeholders	Show the activities required to ensure that the new regime is sustained, supported, and maintained in the long term.

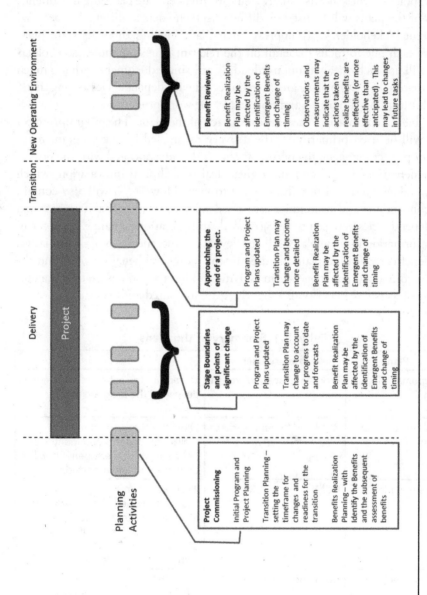

Figure 8.2 A Timeline for Planning Activities throughout the Benefit Life Cycle

The transition plan(s) document the activities required to successfully "go live" and make the change from the old methods of working to a stable new operational environment. This plan will be used by the BCMs to communicate with a wider group of stakeholders in advance of the transition and coordinate the efforts of the transition team and participants. This will cover an intensive period of change activities.

Sustainment plans document the longer-term issues and activities which must be implemented to create and maintain the environment required for the sustainability of the benefits and changes. In some instances, these actions will be addressed at the time of transition, and the sustainment plan will be brief. However, some strategic decisions and actions may be required to ensure the continuity of the changes within the operational environment. For example, there may be a need for changes, beyond the initiative, to sustain the new operational environment, which may be a separate and new program or project.

Figure 8.2 shows the points during the project where planning in relation to benefits may be undertaken:

- Project Commissioning—During the very early stages of the project (often before the project itself has been formally commissioned), project planning is undertaken, and this is the point at which the initial identification of benefits should be conducted. The magnitude and timing of the realization of the benefits should also be estimated.
- Stage Boundaries—Stage boundaries, and the occasions where significant changes are approved, offer opportunities for the project and the benefits to be reviewed and for forecasts to be adjusted based on the most current information available. The things most likely to be re-planned are the timing of the transition and its associated activities and the timing of the benefit reviews.
- Benefit Reviews—During each benefit review, the team measures the benefits realized to that point and confirms whether the forecasts have been met. If the benefits measured are less than forecast, the team may plan additional actions to improve performance to catch up or to modify the timing of future reviews. In the event that a review finds the benefits exceed expectations, adjustments may be made to exploit the good fortune or to identify any emergent benefits.
- End of the Project—Toward the end of the project, the transition plan should be reviewed, and details added to ensure that it is a comprehensive document of all the activities to be undertaken before the outcome can be reached. This plan should reflect the *current* state of the project and its deliverables. The benefits realization plan should also be reviewed to take into account any changes in timing and priorities as a result of the delivery of the project.

8.2.1 Project Planning

When planning a project, a number of process steps must be followed to gather the information necessary to build a comprehensive plan. These steps should be conducted sequentially, because the information developed in one will be a necessary input for the next. In practice, the whole set of steps may become an iterative cycle, with plans being developed and then amended through the next cycle:

- Defining Context
  - Establishing constraints
  - Agreeing on the format and style of the plans
- Defining Scope
  - Gathering information regarding the requirements and content of the project
  - Creating the work breakdown structure for the project
  - Identifying the deliverables and providing a specification for each one
  - Identifying the activities required to deliver these deliverables—including procurement activities if the workload is to be outsourced
  - Identifying the management and quality assurance activities required—these are often overlooked or simplified, which results in insufficient time being allocated for these important tasks or the lack of specialists available to undertake them—which include:
    - Monitoring progress
    - Reporting
    - Quality reviews
- Estimating
  Estimating is a difficult task, but it is one of the most important. It is important because the duration of the project and the funding will be based on the estimates made.
  - Estimates will be required for:
    - Number and type of resources required—to establish the specific skills needed to undertake the workload, the materials, and the funding requirements
    - Duration of each task, considering the number of people available to join the team and suitable as a team size
    - Skills and resources required to undertake governance, planning, and monitoring activities
    - Duration of each of these governance and administrative activities and their impact on the flow of the rest of the project
- Scheduling
  Scheduling is the act of determining the sequence of the activities and their start and finish dates.

- Sequencing—By showing the interrelationships between each activity and others, a network is established which determines the flow of the work involved in completing the project.
- Start and finish dates—Based on the sequencing of the activities and the estimated duration of each task, including the governance and administrative tasks, forecasts can be made of the start and finish dates of each task. By following the sequence established, an overall duration of the project can be calculated. This will set the target dates for the completion of each task, phase, and the project itself, and will determine the milestones which will be reached along the way.
- Assessing Risk
Once the schedule is created, it should be analyzed for risks. Risks manifest as, among other things:
  - A large number of tasks scheduled to start at the same time
  - A large number of concurrent tasks
  - Significant changes in the number of resources required simultaneously
  - The start of one activity that relies on the completion of several others
- Completion and Acceptance
  - Plans should be more than just the schedule of events. They should be complete descriptions of how a project will be managed, controlled, and delivered.
  - Each plan should be completed by recording such important information as:
    - Context of the project and the overall objectives
    - Assumptions
    - Discussions of decisions made during the planning process
    - Constraints and deadlines, or targets set either by the team or by the sponsoring stakeholders
    - Details of the resource requirements
    - Major risks and issues
  - Once the plan is complete, it should be approved by the appropriate parties; generally, this approval will come from the project sponsor or the key stakeholders.

By capturing this information and baselining it, through the approval of the relevant parties, there should be a clear understanding of the project and its goals and objectives to which the key stakeholders agree. This should, at least in theory, avoid misunderstandings and changes of direction later in the project and ensure that the project maintains focus on the primary purpose for which it was established.

The baseline represents the approved version of the plan, or other document, which will be changed only with the approval of an authority figure, usually the project sponsor. The baseline plan represents the current and valid plan, which will be circulated to the team and key stakeholders for reference.

8.2.2 Transition Planning

Transition planning is one area which is too often sacrificed toward the end of a project, when there is pressure to complete on time and within budget. It is, however, a critical area, which can influence the success or failure of an endeavor. Outstanding work in project delivery can be undone, resulting in failure to realize the expected benefits due to breakdown or omission of transition activities.

The BCM must be the person accountable for these activities and time and resources allocated to the transition plan external to the project. This may appear counterintuitive, because there is a project manager and possibly a change manager involved with the project. However, these roles are often assigned to contractors and consultants who are usually outside the sponsoring organization. These candidates for the role will not have the authority and corporate knowledge to identify the need for changes and the impact of those changes within the operational environment. The BCM is already part of the operational team and understands the impact that the transition and outcome will have both on the members of the operational team and on the organization as a whole. The project manager, change manager, and other specialists may be able to advise and support the BCM, but this critical role must be assigned to, and seen to be, an internal appointee.

Transition activities can include many different tasks required as the project draws to an end, and once the project has been completed and its products are handed over to operational groups, end users, and support groups. The transition activities may have a very short duration or may be spread over a significantly longer period. Regardless of the duration, the appropriate resources must be allocated to ensure the successful switch from one method of operating to the new way of working—the new BAU.

Transition activities include the following, although each situation is different, and the level of effort required for each activity may vary:

- Preparation for the change—There is some overlap between communications with the stakeholders and this preparation. It may include amendments to existing processes, the creation of new processes for the new operating environment, and links to other initiatives being undertaken within the organization.
- Training and inductions—There is a need for some training when new systems are introduced, to ensure that all of the people affected by the

change are competent and comfortable with the new systems and can be productive once the change is made. The timing of this training may be difficult to schedule, because undertaking the training too early is likely to be ineffective if the people involved are unable to apply their newly developed skills. Training is most effective if the new skills and knowledge can be used immediately. If large numbers of people require training, this may favor a phased approach to the project and the subsequent changes.

Some of the training may be undertaken while the project is underway or at its completion, during the transition period.

- Assessment of readiness—At some point, the BCM will undertake an assessment of the readiness to implement the change. This will likely be toward the end of the project. As the project nears completion, and its deliverables are handed over to the sponsor, the BCM should review the completeness of the work and ask questions such as:
  o Has all of the scope of the work been delivered?
  o Is everything completed as designed, or have changes been made to the deliverables?
  o Have we received what was expected?
  o Is the operational team ready to take possession and start using the new deliverables and system?

The BCM should understand the situation regarding the feelings of readiness within the internal teams and will make the decision that the preparation is complete, and the team is ready to undertake the transition. Just because the project is finished, this does not mean that the transition *must* happen immediately. As per the example of the new Royal Adelaide Hospital (detailed later in this chapter), if the building and all of its systems are completed during a particularly busy or difficult period of operations at the existing hospital, the best decision may be to delay occupation until a more appropriate time which will result in less disruption to service.

- Achievement of outcome—The outcome is achieved when the new environment is operating. The changes have been made, and the users are working with the new system. In some instances, this will be achieved quickly, and in others it may be at the end of a lengthy phased introduction.

Achievement of the outcome is a significant milestone. It signifies that a new operational state has been reached and stabilized. It demonstrates that the changes planned and required to achieve the benefits have been successfully implemented.

Outcomes, as have been defined, are not the same as benefits; often the outcome is recorded as the changed environment. I often refer to the outcome as the *operational state*—which is one way to delineate between outcomes and benefits. In many cases, the outcome is an important and visible milestone, which should be celebrated. However, this is not the end of the story, and there may be a long gap between the achievement of the outcome and the realization of the benefits. The outcome represents a stepping stone on the way to the longer-term goal. It is likely that some work will be required to maintain focus on the longer-term goals, and additional changes and reinforcement of the change may be necessary at several points.

Outcomes are often important but should not be confused with benefits. The outcome is typically the only visible sign of change that many of the stakeholders will witness for some time. It is important to celebrate the achievement of this goal, which has been the result of significant effort and change. In the case that the project was the development of the capability to enable the online registration of drivers' licenses, applicants and administrators will see that the capability is readily available and working. The immediate uptake of use of the new system may be lower than anticipated—especially if there is an alternative process with which the users are familiar. This makes the immediate measurement of change and benefits subject to unexpected results either significantly higher or lower than the forecast values. As the outcome becomes stable, the measurements will regress toward the mean values—that is, they will stabilize around a level which will be sustainable rather than exhibit extreme results.

- Monitoring and reinforcement of change—There is often a lengthy period of time between the decision to go live and the realization of benefits. This is not always the case, but where it does occur, there is a danger that people may revert to their old behaviors, follow their comfort zone, and use the old system. The outcome may have been achieved initially, but over time there is an attraction to the "good old days" or "good old ways" where the users and stakeholders were comfortable. This is a time of unfamiliarity in many ways; a change has been made, and the team members have been trained and are working differently but observe no benefits or results yet. Once the benefits are realized, they can be announced and the teams know that something has been achieved. However, during this period the stakeholders and users of the new system can become disheartened and begin to look for alternatives.

It is important, during this period, that the users remain focused on the new ways of operating. The BCM must track the mood of the team and ensure that they are happy with the changes and the new methods of working. While "happy" may not be the right term, what must be avoided is the team becoming distressed to the point where they decide to take action.

The BCM should monitor the situation and take appropriate action to rectify any movement away from the preferred new approach. Taking

action covers practically anything that is necessary to maintain the momentum in use of the new methods, and it should be proactive to include:
  - Additional training and induction to ensure that everyone knows what is required, and things that were omitted from or glossed over in the initial training are covered.
  - Reinforcement of the new methods to ensure that the operators understand the importance of persisting with the changes and the new operational environment.
  - Modifications made to the new system. Often, it is only when something is used that errors and enhancements are identified. In these circumstances, especially if the ideas come from the operators, it is important to demonstrate an open mind to the change.
- Decommissioning of existing services and systems—One of the ways to manage the risks associated with switching from one system to another is to have the option of going back (rolling back) to the old system if there are problems with the new system. Once the outcome has been achieved, and the new system is stable in operations, the BCM will make a decision that a rollback to the old system is no longer feasible or even possible. In the event of a team moving from one building to another, it would not be possible to "roll back" to the previous office once it is leased, sold, or occupied by another group.

In many cases, having two systems available may lead to the use of the superseded one, either by mistake or preference. As soon as the new working environment has been deemed stable and reliable, the BCM should make the decision to decommission the existing systems. This decision is based on the operational environment, so the decision that this point has been reached should be made by the person with the greatest operational knowledge and responsibility—the BCM.

Once the decision is made, arrangements can be made for decommissioning the systems which are no longer required. Decommissioning should be seen as an all-inclusive term, which may cover:
  - Archiving and deleting/destroying the old documentation, processes, and software
  - Removing artefacts and templates
  - Sale of assets that are no longer required
  - Demolition of buildings (which is a large project in itself)
  - Dismantling of oil rigs and power-generating plants

Significant effort and resources will be required to achieve some of these decommissioning activities. These costs and effort must be planned for and accounted for in the business case. Failure to include these costs in the business case will result in poor and ill-informed decisions being made.

8.2.3 Post-Transition—After the Outcome Has Been Realized

Once the transition has been completed and the outcome has been reached, there will be a period when the new operating environment is stable. In some cases, this is the point at which benefits will start to be visible and measurable. In other cases there will be a lag between the end of transition (when the new methods of working become the standard operating environment), and the benefits are being measured.

Often there is a need for additional work to ensure that the benefits are generated. In other words, the benefits do not always *magically* appear. They need to be realized through hard work during the delivery stages of the program, the transition phase, and after the transition is completed. Clearly, this post-transition workload needs to be planned and coordinated, and a number of activities need to be planned during this phase.

- Benefits reviews—The benefits need to be measured. This requires the cooperation of the BCM and operational personnel in addition to those who undertake the reviews. Generally, these are relatively short assignments to record the realized value of the benefits at a point in time and the collation of supporting evidence. These reviews should be conducted by independent assessors and planned to optimize these resources.
- Communication of results and progress—A common issue with changes initiatives in organizations is that many people are affected by the disruption of the change, but few receive any commentary or results about that change. That is, the actual benefits are not widely publicized. Some stakeholders will not be aware of the success of the endeavor and will regularly be asking, "Was it worth it?"

    There will be a need to report the results of reviews to the sponsor, in particular, and those with an interest in the business case and investments. However, the other users and other stakeholders should receive information regarding the benefits as a way of maintaining their interest and generally keeping them informed. Clearly, this is easier when the benefits are as forecast or exceed forecasts. In the event that the measured benefits are lower than expectations, it is also important to explain why that is the case, and what (if any) responses are being implemented to address this issue of underperformance.
- Decommissioning of defunct systems—Decommissioning is usually associated with the transition. However, there may be circumstances when the decommissioning of the obsolete system must wait until a later time. The decision that the old system is no longer required must be made from an

operational perspective, and therefore, the BCM is the person most likely to make an informed decision regarding the readiness for and timing of the decommissioning.

Depending on the overall initiative and the projects completed, the decommissioning may be a simple task or may be a complex project in its own right. Regardless, the work must be adequately planned, with key milestones and decision points clearly identified.

- Responsive actions—As the progress toward benefits realization advances, new information may come to light, especially for the BCM. This information will inform decisions to take additional actions in response to the success of the program to that point. Three general actions may be appropriate:
  - Reinforcing actions—Required to reinforce the new way of working and avoid any slippage back to previous work methods. This may take many forms, such as additional training and setting performance goals.
  - Corrective actions—Required if the measured results do not match the forecasts and there is a need to respond.
  - Scope changes—Information gleaned from the progress of the work and increased familiarity with the work and change may indicate that a change of scope is advantageous. This might be a reduction in the scope of the work or cancellation of a project, because the benefits measured at that point are greater than expected and the original scope is not required in its entirety to achieve the objectives. Alternatively, there may be an opportunity to increase the scope and projects to take advantage of an opportunity or some emergent benefits.

8.3 Planning for Benefits

The planning required for benefits realization is complicated by the fact that the benefits will be spread over an extended period, often years after the completion of the project, and that there are, generally, a greater number of stakeholders to be consulted. There will also be intensive periods of planning and implementation activities at different stages of the project.

Figure 8.2 shows a single project as a solid block of activity and progress. The planning for benefits needs to include:

- Identification and analysis of benefits—This is often perceived as part of establishing the context for the project and its associated changes. As such, it is not always included as activities within the plan, because it will be completed before the project team is assembled. This is a factor of the maturity of the organization with respect to projects and programs and its existing corporate processes. So, it may be seen to be a part of gaining approval for the project rather than a project-related activity. It is vital, however, that sufficient resources be committed to this activity to ensure that there is a common and detailed understanding of the purpose of the project.

- Communication activities—As the project starts and people can see that something is happening, there will be need to explain the changes and the impact they will have on the users and others. In fact, one of the most important facets of the relationship between the stakeholders is trust. To establish that trust, it is wise to get the message out to the stakeholder community as early as possible, and certainly before the information emerges from another source.

  At the earliest opportunity, the BCM should plan to hold information events, even if simply to announce that the project has beem proposed and will be explained at a later date. Obviously, the amount of information which can be released will depend on a number of factors, including:
    o Contract constraints
    o Confidentiality and the security of data
    o Sensitivity of the information
    o The impact of the early release of information
    o Completeness and reliability of information

  A number of the stakeholders of a program which will result in users of benefits will not be involved directly in the delivery process. It is important that they are provided with sufficient information regarding the program and the subsequent change and especially how that will impact them directly. Failure to engage these stakeholders may result in delays to the project, anxiety among this important group, interference with the progress of the change, and, in extreme situations, sabotage and obstruction.

  These communication events are often presented and undertaken by the program team or the sponsor. In some cases, it may be advantageous for these events to be presented by the BCM or another party who is seen to be more empathetic with the broader stakeholder community. The BCM should be appointed from an operational role and as such is viewed as being on the side of the stakeholders, which is very different from the perception of the "project team" as being on the side of the organization. It is vital to the overall success of the initiative that change is led by a credible authority. Having BCMs who lead the areas which need to undergo the changes ensures that a recognized leader, who is accepted by the users, is in a position to represent operational interests. This internal appointment is more likely to have empathy with the affected users and will have the authority to make decisions regarding the implementation of the new systems and ideas.

The communication events should be planned taking into account the availability of relevant information, the right people to present the information, and the appropriate point in the progress of the work to release information.

From the BCM's perspective, these events should include:
- Announcements of major decisions and achievements
- Training and induction of stakeholders
- Discussions of the impact

• Review points within each project—There is an expectation that the BCMs will be kept informed of the progress within the projects which impacts their operational area. Although the BCM and the stakeholders he or she represents may have no active role to play within the project and may undertake no project-related/specific work, the BCM does need to be familiar with the progress in order to adjust the timing of messages to the stakeholders. The BCM should schedule time to receive regular updates and briefings from the project manager. This may be in the form of attendance at regular progress meetings, individual briefings, and inclusion at reviews at milestones in the delivery of the project (project management life cycle). The timing, relevance, and inclusion in regular meetings will depend heavily on the project delivery life cycle favored by the organization.

The BCM should base the planning of benefits on the information available from within the project. Primarily this will inform the planning of communication events, training, and transition activities. As the project progresses, it is likely that changes will be made to the project schedule and to the scope and content of the project. These changes may affect the message passed on to the stakeholders to ensure that they are kept informed accurately and understand how the change will affect them. For example, if construction of a new office is behind schedule, early notification to the incoming occupants will assist with their planning and preparation and may allow the occupants to plan their relocation differently.

The New Royal Adelaide Hospital is a large new hospital development that is instrumental to the health services in the state of South Australia. At $2 billion, this is one of the largest hospitals ever constructed, and the magnitude and complexity of the work have resulted in delays to the construction process. This will obviously affect the timing of the relocation of many of the services and personnel. With regular progress information being available, the current units can modify the timings of their relocation and induction activities to suit them. However, the timing of some of the transfers is further complicated by the fact that the availability of the new facility is scheduled to coincide with an extremely busy "season" in some of these departments, and it would not be sensible to relocate during that time of the year. Unfortunately, this is likely to lead to a delay in achievement of the complete outcome (a functioning hospital facility), and there will be a delay in the realization of some of the benefits.

As in this example, the transition activities may be delayed, or brought forward, based on the progress of the project. This will result in replanning those activities, and the need to send out different, amended messages to the stakeholders affected. In addition, the timing of the realization of the benefits may be affected if the project is delivered earlier or later than scheduled. This may seem like a relatively minor inconvenience, but delaying transition and benefits realization may move the benefits into a different fiscal period, perhaps diluting their value. This seemingly relatively small change (perhaps the discount rate for calculating the net present value is 5%) means this reduction in the quantity of a benefit may have been forecast as $1 million in, say, 2018, but if delayed by several months the benefits may not be realized until the following fiscal year (2019), and the net present value of those benefits will be reduced by roughly $48,000. The timing of subsequent benefits may also be affected in a domino effect, which will delay and reduce the return on the investment. The initial investment may not have altered, with payments for the project being processed on schedule during 2015 and 2016, but this will exaggerate the gap between the investment and the return quite quickly and reduce the viability of the overall business case.

8.3.1 Early Wins

Once a benefits realization plan has been developed, reviewing it and revising it can reap significant rewards. Early positive results can reinforce the need for the investment and motivate the team and users.

"Early wins" refer to the achievement of early positive results, which can sometimes be planned into the initiative, ensuring that the stakeholders observe a visible and effective change having a real impact on the organization and the broader stakeholder community.

There are two reasons for searching for early wins and changing the plan to take advantage of them:

- Stakeholder engagement
- Improving the business case

Stakeholder Engagement—Enhancing the Stakeholder Experience

One of the problems with programs is the lag between visible activity and results. Programs and projects commence with a great deal of activity, some of which is visible to the stakeholders, while other work may be hidden from sight because it is conducted off-site. While stakeholders may appreciate that work has started, they are also impatient for results. The gap between the work (particularly the start of the work) of the initiatives and the changes in the operational environment may be considerable. The lag between the changes in that operational environment and the benefits being realized and publicized may be considerably longer. This can be disappointing and discouraging for the stakeholders, resulting in a loss of interest in the initiative. To maintain that interest, which will be valuable when the stakeholders' support is needed, the team should review the plan with the aim of bringing forward the realization of some of the benefits so that the gap between action within the projects to the change activities and to the realization of benefits is reduced. Stakeholders will then observe results more quickly.

When managing a program, this could be achieved by starting some of the smaller (shorter-duration) projects earlier, so that they are undertaken, completed, and achieve their results early—even if the value of those benefits is not high. Figure 8.3a shows the original schedule/timeframe for delivering an initiative with six projects. Figure 8.3b shows a modified timeline in which three of the projects (3, 5, and 6) are delivered earlier, and as a result, the associated benefits are generated earlier than with the original schedule. The visibility of the early results is important in maintaining interest among the stakeholders.

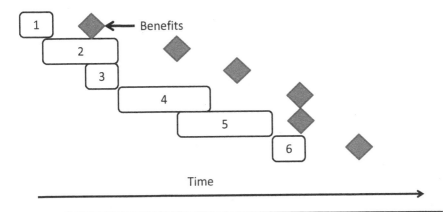

Figure 8.3a The Concept of Early Wins—Original Schedule of Projects and Benefits

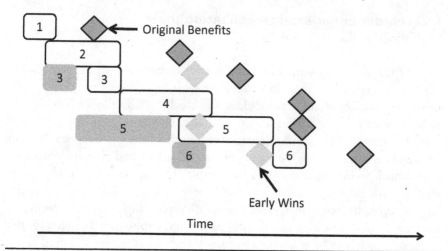

Figure 8.3b The Concept of Early Wins—Adjusted Schedule to Realize Benefits Earlier

Additionally, the communication involved in developing the plan may identify some existing issues which are a nuisance to the operational environment. It may be possible to resolve some of these issues during some of the early projects. The stakeholders will see that the program has a direct, positive impact on them and acknowledge that their needs are being addressed. These results may be small in magnitude, but because they are delivered early in the benefits life cycle, they can contribute to enhancing the viability of the business case. In many cases, it is often more important that these wins are visible, they are reported widely to create enthusiasm for the change, and they serve as a case study for use in promoting the initiative's success among the stakeholders.

Improving the Business Case

The early realization of benefits is likely to have an impact on the business case for the program. In some instances, it will be possible to rearrange the timing, and possibly the sequence, of projects within a program at little or no extra cost. If the resources are available, and there are no dependencies, the early delivery of projects will result in the early realization of benefits. This will increase the strength of the business case.

In addition, the net cash flow will be improved. This is particularly important if the sponsor is borrowing the funds for the program, because it will result in the need to borrow less, as the early benefits generated will offset the costs of delivery.

Table 8.2 Data for Five Projects within a Program

Project	Cost	Duration	Benefits	Period of Realization	Lag between End of Project and Benefit Realization
A	$3,000,000	3 years	$4,000,000	4 years	2 years
B	$2,000,000	4 years	$2,500,000	5 years	1 year
C	$1,000,000	1 year	$1,500,000	3 years	2 years
D	$2,000,000	4 years	$2,000,000	2 years	3 years
E	$500,000	1 year	$750,000	1 year	0 years

Table 8.2 shows data about an initiative and the timing of projects and benefits. It assumes that the costs of the project are spread equally over the period of delivery, and the benefits are realized equally over the period of realization. Figure 8.4a is a spreadsheet showing the initial plan for this program and the annual costs and benefits. Of particular interest in the discussion of early wins is the Net Cash Flow line. This peaks in Year 5 of the program. These figures are not discounted so do not represent the NPV figures. Finally, Figure 8.4b shows the impact of realigning the projects. In this example, Projects C and E are scheduled to start in Year 1.

In this example, the overall position at the end of Year 11 is the same, but bringing the two projects forward reduces the lowest point for the net cash flow position (and greatest need for financing) in Year 5 from $7,500,000 to $6,250,000. The net cash flow is reduced because the early realization of benefits offsets some of the expenditure. The impact would be greater in a longer-term program or if net present value was applied. However, it is clear that revising the plan to take into account early wins can impact the business case.

There is another consideration with respect to the business case of a program or concurrent projects. The following example is adapted from an article published, on LinkedIn, by Bill Duncan (2015).

Consider that an organization needs to undertake two projects, which have no technical overlaps or dependencies. Both projects can be conducted by the same 10-person team, and each has a budget of $500,000 and a timeframe of five months, with an even distribution of costs over the project duration, that is, $100,000 per month.

The organization wants the projects to be undertaken concurrently. The team of 10 could be split in two, and a team of five applied to each project. This results in an adjustment to the schedules, with each project having a duration of 10 months, with no change to the budget. Assume all of the estimating, scheduling, and development of a budget is accurate. The risk profile for each project is similar, and these risks are controlled.

Project	YEAR										
	1	2	3	4	5	6	7	8	9	10	11
A	1,000,000	1,000,000	1,000,000								
B	500,000	500,000	500,000	500,000		1,000,000	1,000,000	1,000,000	1,000,000		
C						500,000	500,000	500,000	500,000	500,000	
D			500,000	500,000	1,000,000	500,000		500,000	500,000	500,000	1,000,000
E					500,000	500,000	750,000				
COSTS	1,500,000	1,500,000	2,000,000	1,000,000	1,500,000	1,000,000					
BENEFITS						1,500,000	2,250,000	2,000,000	2,000,000	2,000,000	1,000,000
NET CASH FLOW	-1,500,000	-3,000,000	-5,000,000	-6,000,000	-7,500,000	-7,000,000	-4,750,000	-2,750,000	-750,000	1,250,000	2,250,000

Legend: ☐ 500,000 Project Cost ■ 500,000 Benefit

Figure 8.4a Timeline of the Data in Table 8.2

Project	1	2	3	4	5	6	7	8	9	10	11
A	1,000,000	1,000,000	1,000,000			1,000,000	1,000,000	1,000,000	1,000,000		
B	500,000	500,000	500,000	500,000		500,000	500,000	500,000	500,000	500,000	
C	1,000,000			500,000	500,000	500,000					
D			500,000	500,000	500,000	500,000				1,000,000	1,000,000
E	500,000	750,000									
COSTS	3,000,000	1,500,000	2,000,000	1,000,000	500,000	500,000					
BENEFITS		750,000		500,000	500,000	2,000,000	1,500,000	1,500,000	1,500,000	1,500,000	1,000,000
NET CASH FLOW	-3,000,000	-3,750,000	-5,750,000	-6,250,000	-6,250,000	-4,750,000	-3,250,000	-1,750,000	-250,000	1,250,000	2,250,000

Legend: □ 500,000 Project Cost ■ 500,000 Benefit

Figure 8.4b Timeline for the Projects Seeking Early Wins

Table 8.3 Two Concurrent Projects

Month	Concurrent Projects			
	Project A		Project B	
	Costs	Benefits	Costs	Benefits
1	$50,000		$50,000	
2	$50,000		$50,000	
3	$50,000		$50,000	
4	$50,000		$50,000	
5	$50,000		$50,000	
6	$50,000		$50,000	
7	$50,000		$50,000	
8	$50,000		$50,000	
9	$50,000		$50,000	
10	$50,000		$50,000	
11		$60,000		$60,000
12		$60,000		$60,000
TOTAL	$500,000	$120,000	$500,000	$120,000
Combined Total:				
	Costs	$1,000,000		
	Benefits	$240,000		

Upon completion, it is expected that the projects will return a benefit of $120,000 ($60,000 from each project) per month through cost savings and efficiencies. Starting the two projects on January 2 (allowing for New Year's festivities) and completing the two projects on October 31 would seem a good result.

Table 8.3 shows that the expenditure over the course of that year will be $1,000,000 spread over the first 10 months, followed by the realization of $240,000 worth of benefits over the final two months of the year. The net position is that the organization has spent $760,000. Presumably there would be more benefits to follow in subsequent years.

Is there a better way of planning this work? What would happen if the organization delivered one project at a time? In that case, the whole team would be able to focus on one project and deliver it in half the scheduled time. The organization would then be able to realize the benefits from that first project while working on the second project. If both projects are delivered on schedule, there will be a period when benefits are being realized while the second project is underway.

Table 8.4 shows that the effect of focusing on one project at a time. Project A would cost $500,000 and be completed by the end of May (five months). At this point the completion of the first project would generate benefits of the value of $60,000 per month. Over the course of that year, that is, the remaining seven months, this would amount to a total of $420,000 in benefits. From June to October, the second project would be undertaken, again at a cost of $500,000, making the total Year 1 investment $1,000,000. The second project would generate two months' worth of benefits, totaling $120,000. The overall position would be an outgo of $1,000,000, offset by the realization of benefits totaling $540,000 ($420,000 from Project A plus $120,000 from Project B). The net investment during Year 1 would therefore be $460,000. Compared to undertaking the two projects simultaneously, this is better result to the value of $300,000 during the first year and would reduce the payback period.

This example is somewhat idealistic and may be difficult to replicate in reality. However, the advantages of bringing forward the benefit realization phase

Table 8.4 Two Simultaneous Projects

Month	Concurrent Projects			
	Project A		Project B	
	Costs	Benefits	Costs	Benefits
1	$100,000			
2	$100,000			
3	$100,000			
4	$100,000			
5	$100,000			
6		$60,000	$100,000	
7		$60,000	$100,000	
8		$60,000	$100,000	
9		$60,000	$100,000	
10		$60,000	$100,000	
11		$60,000		$60,000
12		$60,000		$60,000
TOTAL	$500,000	$420,000	$500,000	$120,000
Combined Total:				
	Costs	$1,000,000		
	Benefits	$540,000		

is clear and may be financially significant. Early wins can change the viability of a program or project and provide comfort to the stakeholders.

Measurement and Review of Benefits

At some point, or points, after the outcome has been achieved, the benefits need to be measured. A review comparing the actual results against the forecasts will also be undertaken. There may be logical points in the project and change during the life cycle when it is obvious that these measurements should be conducted. In other instances, this may not be the case, and a more sporadic approach is taken to the timing of measuring the benefits.

Regardless, there should be a plan which communicates the timing of benefits measurement activities, reporting, and the taking of corrective or proactive actions. Measuring benefits will take effort, and resources will be applied to these activities. Planning will certainly assist with understanding the impact of events within the project, including risks. If the project completion is delayed, it is likely that the outcomes and benefits will also be delayed, possibly pushing these achievements into different reporting periods (e.g., financial years), which may have a detrimental effect on the business case.

Some considerations should be taken into account when scheduling these important activities:

- Integration—Can existing reporting mechanisms be used to address the benefits and progress? If there are already processes and reporting requirements within an organization, it will be an advantage to integrate the benefits realization reporting into the existing mechanisms. This will reduce effort, particularly in reporting, and help to embed benefits management into the culture of the organization.
- Information contribution—Are specific reports necessary, such as does the project sponsor need to report upward? In this case, the timing of the benefits review and the reporting of it will be important, as there will be an expectation that the information will be made available in time for the sponsor to report accurately.
- Resourcing—Measuring the benefits will take time and effort. Resources will need to be identified and deployed at the appropriate time.
- Consolidation—It may be possible to measure several benefits at the same time and attribute them to the whole program. In projects and change initiatives where a number of benefits can be measured simultaneously, it may be convenient to consolidate a number of benefits to be measured at the same time. This may simplify matters in terms of reporting and make

the process less labor-intensive; it may even reduce the costs of measuring the benefits.
- Assurance and independence—In many ways, the obvious role to be responsible for the measurement of the benefits is the BCM. However, the BCM may be too close to the change and may have an interest in the value of the benefits reported. It may be difficult for this person to be or be seen to be unbiased. It is recommended that the measurement of benefits be conducted by an independent person or group; they can be from within the organization but should not have an interest in the project or the change. Likely sources of candidates are:
  - Portfolio management office (if one exists)
  - Program management office
  - Subject-matter experts from other parts of the organization
  - Independent contractors/consultants—The use of specialists who are independent of the project and the change is important in giving the project sponsor confidence that the figures are accurate and reliable. This extra layer of project assurance will confirm the values and provide the information needed to verify the business case.

Front-End Loading

Front-end loading (FEL) has been mentioned already, but its significance in the implementation of programs, projects, and change initiatives should not be underestimated. FEL is the approach of allocating significant resources to the early phases of the initiative to ensure that the scoping, planning, and definition activities can be undertaken thoroughly. Time and resources are specifically set aside for these activities.

Once a plan is developed, it should always be reviewed and refined to optimize the delivery of the projects and the allocation of resources. With respect to the realization of benefits, the complete life cycle can be optimized, including the delivery of the projects, the efforts required for a smooth transition, and the timing of the benefits. Being able to "bring forward" the realization of benefits can have a major impact on the program from two aspects, as discussed previously. First, they can be used to promote the progress of the initiative by reinforcing the achievement of milestones and the goals; and second, the early realization of benefits can change the net present cash flow of the project and potentially its business case.

The development and implementation of plans enable the team to deliver its work in an efficient and productive manner. However, the time and resources required to create useful and robust plans are not always forthcoming. In part this is because of the confusion between *productivity* and *busyness*. Projects are often forced to begin before adequate planning has been completed because stakeholders want to see the team active. Action is often more sought after than thought. It is a compelling argument that the program or project has just been approved, and the stakeholders are expecting progress. Stakeholders have just come to an agreement about the project and its objectives. Edkins et al. (2013) found that the role of the sponsor and other key stakeholders was critical to the success of the work conducted at the front end of the project. Although there was some disagreement about what constituted the "front end," the evidence of structured management and project management processes to this phase established the importance of this period. Edkins et al. (2013) concluded that the role of the project manager during this period should focus more on challenging the viability of the investment.

Assembling a team for the purpose of thinking about the initiative for weeks or months is not necessarily the response stakeholders want to see. It is, however, the smart thing to do. Merrow (2011) found that megaprojects which invested heavily in time and people during the initial stages were four times more likely to be successful. A relatively small commitment (perhaps 3–5% of the total cost) devoted to planning and preparation will lead to reduced risks and greater certainty. It is a great investment in the initiative.

8.4 Documentation

8.4.1 Program Plan

The *program plan* is the key reference document for the initiative. It contains the schedule, resource, and cost information. Additionally, it should contain or refer to all of the key information required for the delivery of the program, including scope, team and roles and responsibilities, controls, and plans for engagement of stakeholders. It should be updated as the work progresses. One way of considering the program plan is that it will be used to induct new team members and needs to contain all of the program information they will require.

Prepared by the program manager with input from project managers and others, the program plan includes information regarding:

- Overall schedule for the initiative
- Assumptions, constraints, and lessons which were applied in developing the plan
- Explanation of the projects selected and their groupings and other activities

- References to major risks and issues and direction regarding their management (which may take the form of a risk management plan and issue/change management plan)
- Dependencies between projects and activities
- External dependencies and decisions from outside the initiative
- Major milestones for the progress of the initiative
- Direction regarding the monitoring and reporting of the advancement of the delivery of the initiative
- Costs and resource requirements for the delivery of the plan

8.4.2 Benefit Realization Plan

The *benefit realization plan* is more than just the schedule of points at which benefits will be realized and can be measured. That information is part of the plan, but additional detail will be included, including:

- Who will be involved in the identification and definition of the benefits
- What are the points at which the plans and business case document will be reviewed
- What are the details of the methods to be applied, and their timing, for reporting
- What resources are required to measure the benefits
- What are the details of the relationships between the benefits and their connections to the projects
- What are the details of individual benefits (that is, the benefit profiles)

If the benefit realization strategy provides the guidelines and rules for managing benefits, the benefit realization plan is a live document, which provides the details of how those guidelines will be applied for the initiative.

It is tempting to merge the benefit realization plan with the project management plan. This should be avoided for two reasons. First, there will be more project-related tasks and planning documentation than benefits documentation, so the benefit realization plan will become a minor section of the larger project-focused document. Benefits management will become marginalized and possibly forgotten in the need to focus on the project-related elements of the change. Second, the benefit realization plan will extend beyond the project. Once the project is completed, and its documentation archived, there will still be a need for a reference document for the details of the ongoing measurement reviews and reporting for the benefits. This is best achieved by keeping the project planning documentation and the benefits planning documentation as separate but linked, so that if one set changes, the other may be adjusted for consistency and currency.

8.4.3 Transition Plan

Several *transition plans* will be required, one for each of the transitions which take place during the initiative. Each transition plan details all of the activities, resources, and costs associated with the work required when the outputs from projects are integrated into the operational environment. It will be used predominantly by the business change manager to prepare for the communication of the changes and the delegation of work to all of those involved in the transition period. Some of the activities in this plan may overlap with the project, and key information is the dependencies between the project and transition.

Each transition plan should discuss:

- Triggers for the plan to be enacted
- Dependencies between transitions and projects
- Constraints which apply to the transition activities
- Overall schedule for the transition
- Assumptions and lessons which were applied in developing the plan
- References to major risks and issues
- Costs and resource requirements for the delivery of the plan

8.4.4 Sustainment Plan

Some activities may be required to be undertaken in order for the changes enacted and the benefits to continue. This work may be required outside of the initiative and could involve corporate groups or functions. The *sustainment plan* is developed by the program manager to document these requirements with timeframes and a clear discussion of the connection between the continued realization of benefits and these actions. The plan will need the input of those groups who will need to implement the sustainment actions.

The detail within the sustainment plan will be progressively elaborated as the initiative advances and will be handed over to the sponsor to be allocated to the appropriate parties for action and consideration. As a minimum, the sustainment plan should contain:

- Triggers for the plan to be enacted

- Assumptions and lessons which were applied in developing the plan
- Reasoning behind the need for the sustainment activities
- Dependencies between sustainment activities, the benefits, and corporate objectives
- Constraints which apply to the sustainment activities
- Overall schedule for the implementation of the sustainment plan
- Explanation of the impact of the "do nothing" scenario
- Costs and resource requirements for the delivery of the plan

8.5 Summary

Planning for benefit realization is not a simple task. It must consider the delivery aspects and the views of the stakeholders, particularly the users. These stakeholders, in particular, will be those most immediately affected by the changes and transition. The planning of activities relating to the management of benefits must be integrated into the project planning and the dependencies between the different plans clearly identified. Generally, four plans are produced which are relevant to benefits realization management:

- Program/project plan
- Benefits realization plan
- Transition plan
- Sustainment plan

The plans will not remain static documents for long, and there must be a commitment to maintain the plans incorporating the deviations from them and changes which have been approved. Once these plans are updated, it is important that changes are communicated to the relevant stakeholders to ensure that expectations are managed and there are no surprises.

Table 8.5 summarizes the activities conducted within the process of planning the benefits.

The following chapter will discuss the coordination of the workload and tasks involved in realizing the benefits. Some of the issues addressed during this process should be

- Are the resource requirements understood?
- Have milestones been set for the touch points between the plans?
- Have communication links been established to coordinate the work and the changes to the plans?

Table 8.5 Activities Conducted within the Plan the Benefits Process

Activity	Purpose	Responsibility	Documentation
Workshops and subsequent planning	Develop the program plan	Program manager (coordinating the team)	Program plan
Workshops and subsequent planning	Develop transition plans	BCMs and change teams—the program manager prepares these plans (PMI, 2017b)	Transition plan(s)
Workshops and subsequent planning	Develop sustainment plans	Program manager and sponsor	Sustainment plan
Review and revise benefit profiles	Update the information regarding each benefit, if the attributes have been changed during the planning process	Program manager and BCMs	Updated benefit profiles
Review and business case	Update the business case document	Program manager	Updated business case document
Approve the initiative plans	Confirm that the business case is acceptable to the key stakeholders	Sponsor and sponsoring group	

Exercises and Activities

Table 8.6 contains the costs and benefits from the initial plan of an initiative comprising five projects. Project costs are shown in black, benefits are shown in **BOLD**. Costs and benefits are in $000. Assume that all projects are independent.

Table 8.6 Timeframe for Five Projects

Project	Year									
	1	2	3	4	5	6	7	8	9	10
A	1,000	1,000			**750**	**750**	**500**	**500**		
B		500	500	600	**450**	**300**				
C		250	250	250	250			**300**	**400**	**500**
D			300	750	500		100	**600**	**1,000**	
E			100	300	250	250	600	**250**	**250**	

1. Rearrange the timing of the projects to enhance the early wins. Take into account the following constraints:
   a. A maximum of two projects can be started in the same year.
   b. The sponsoring group can release a maximum of $2,000,000 in any year.
2. Is it possible to adjust the schedule so that the worst net cash position is below $4,000,000?
3. What front-end loading activities could be undertaken in your organization's program environment?
4. Consider a program within your organization or with which you are familiar. What sustainment activities would be required to establish a long-term environment in which the benefits will continue to be realized?

Chapter 9

Coordinate and Realize the Benefits

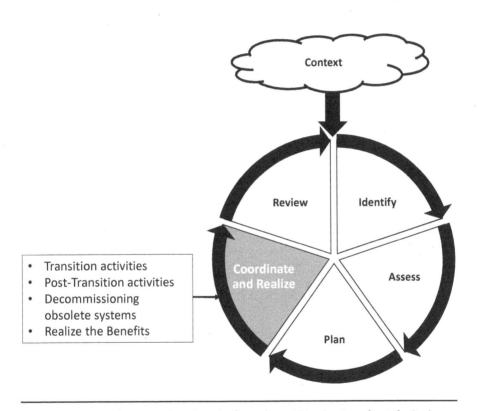

Figure 9.1 Coordinate and Realize the Benefits within the Benefits Life Cycle

"You can create value with breakthrough innovation, incremental refinement, or complex coordination. Great companies often do two of these. The very best companies do all three."
— Sam Altman

"Music is . . . the coordination between man and time."
— Igor Stravinsky

"If everyone is moving forward together then success takes care of itself."
— Henry Ford

Coordinating and realizing the benefits is one of the most significant and influential phases of the benefits life cycle. Figure 9.1 shows the process in context which is applied as the project(s) are completed. Products and outputs have been delivered, but now they must be put to work. It is only through the successful integration of these new products into the operational environment that there will be outcomes and then benefits. A project delivered on time and within budget is important. However, the return on the investment comes from the realization of benefits.

The work involved in coordinating and realizing the benefits should be led by the business change managers (BCMs). This is another reason for that role to be appointed from an operational position of authority. Once one of the projects is completed, there will be a period of transition during which the operational environment will take ownership of the products and systems produced. This effort establishes the new methods of working. Following the successful transition, these new methods of working must be sustained and supported to generate the benefits. Both of these activities need leadership from somebody in authority in the operational environment, that is, a functional manager.

This chapter will explore the ways in which BCMs can influence the success of the initiative through their participation during the delivery of the projects and their specific role in the post-project period when transition takes place and benefits are realized.

The BCM's role moves through three distinct phases:

- Pre-transition—When the planning and preparation for the transition are undertaken
- Transition—When the activities to accept the products from the project and to integrate those products into the operational environment take place and result in the outcome being achieved
- Post-transition—When the benefits are realized and managed

All three phases are important and must be undertaken by the BCMs utilizing their support to ensure the management of a smooth pathway from the project life cycle into the operational world. Once the outcomes have been achieved, the benefits should follow, some additional activities are likely to be required to maintain the momentum toward the ultimate goal.

9.1 Pre-Transition

Pre-transition activities include all the planning that is required for the transition as well as a significant amount of communication with the stakeholders in the operational team. Merrow (2011) and Edkins et al. (2013) emphasized the importance of committing time and resources to these preparatory activities, which could include training of personnel, modifications to existing systems, and coordination of the timing of transitions. In a program environment, where multiple projects are undertaken concurrently, the result could be that a number of projects are completed at the same time. Care should be taken in this instance because the result is that a number of projects will go through their transition and go live at the same time. The operational environment, the personnel, and potentially other stakeholders, may not be ready for such significant or numerous changes all enacted at the same time. Therefore, one of the roles of the BCM is to ensure that the operational environment is ready to accept new products and the associated changes. The decision to go live must be made by somebody with operational authority and responsibility—the BCM.

9.1.1 Changes to the Project

While all of this action is taking place, the BCMs should be monitoring the project for changes. A number of people will already be monitoring changes within the project, so why do the BCMs have this additional burden thrust on them?

Traditionally, change control within a project has focused on how the adjustment of scope impacts the other two of the Triple Constraints, that is, how the project's schedule and budget will be impacted by the changes once they are approved. The BCM needs to look longer-term and understand how the changes will impact the transition and realization of the benefits. The BCM will need to make the case for changes which could improve the business case through the enhancement or optimization of the benefits. The BCM may also need to argue against ideas for change which devalue the business case or make the transition more difficult. Although the management and ownership of the business case, rests ultimately with the sponsor, the BCMs are the closest stakeholders to the operational environment and should be able to forecast how the changes will affect the teams during transition and beyond, when the benefits will be realized.

Ultimately, decisions about change will be made by the sponsor, but the BCMs should expect to be part of the discussion when these changes are considered and the change control processes are applied.

9.1.2 Scope Changes

There are a number of reasons for considering changes to the scope of a project, including:

- Technical difficulties regarding the production of products
- Completion of the original scope may cause unacceptable delays
- New technologies may enable the increase of the scope to cover additional functions and features
- Changes to the operational environment

Each of these will have an impact on the transition and operational environment by changing the requirements for training and familiarization with the new system, the capability and operational environment itself, and the benefits which can be expected because of the new features.

In some cases, such as the decision to de-scope, or remove functions and features form the scope, there may be a reduction in the overall benefits which will be realized. This may occur when the scope is reduced to meet deadlines or budgetary constraints. Although money may be saved in the project, the business case will also be impacted because of the removal of the functions.

Sometimes, a change may result in the expansion of the scope to include features which were not considered previously. These may be considered because new technologies are available, cost-effectively, which were not available earlier. This may cause the business case to be revised because changes should be considered only when they add value to the investment. The BCM will need to revise the plans and transition activities to take into account the changes.

9.1.3 Schedule Changes

As the initiative progresses, changes will be made to the schedules. Regardless of the reason for deviating from the original plan, any changes to schedule will impact the transition and timing of the realization of benefits. The BCM must understand how these schedule changes will affect the transition plan and the subsequent actions required to realize the benefits. The BCM may need to change the communication plans to advise stakeholders of the changes and to modify future communication and engagement events.

Particularly when the program will result in significant change, there may be apprehension and anxiety among the stakeholders. Surprises and sudden changes, no matter how well intentioned or advantageous, are likely to result in negative feelings among the stakeholders. This may lead to a less efficient and less productive transition and take-up of the new systems, resulting in a delay in achieving the outcome.

9.1.4 Training

Some specific training may be required in preparation for the transition activities and to achieve the outcome. This will need to be planned and resourced. To ensure that the training is effective, it will be necessary to undertake it close to the time when the newly developed skills can be applied. A lengthy gap between the training and starting to use the learned skills may make the switch to the new methods of working less effective and may lead to the need for further reinforcement training at a later date. This additional training is expensive and it directly impacts the productive working of the teams involved. The need for additional training is likely to be identified only after a period of unexpectedly low productivity, or lower than forecast benefits. This means that ineffective training may result in the following dis-benefits:

- Lower performance during the transition period
- Reduced performance, compared to forecasts, following achievement of the outcome
- Additional training costs
- Reduced performance, or increased disruption, during the period of additional training
- Reduced morale within the team

The individual, and cumulative, impact of these results will reduce the viability of the business case by raising costs and/or delaying the realization of benefits.

Training, regardless of when it is planned, will be disruptive to the operational teams. The BCM needs to monitor the progress of the projects to schedule the training to fit in with operational needs and commitments.

9.1.5 Communications

In addition to training, there will be a need to make the stakeholders aware of the changes and how each group of interested parties will be affected by the transition and changes in the environment. The BCMs will be able to advise the broader stakeholder groups with authority.

There should be a consistent source of information which can be relied on. In many cases, the communications will provide updates and revisions to keep the stakeholders informed of progress. The BCMs and other parties who will be involved in the communication should coordinate the messages to cover:

- The need for the change
- The expected outcomes and benefits
- The contribution stakeholders can make to the initiative
- The impact of the change on stakeholder groups
- Changes to the existing environment
- Revisions of the plans
- Reinforcement of the need for change

9.1.6 Baseline

A baseline needs to be established to measure the existing "normal" levels of performance so that any increased performance can be measured and compared. The need to establish a reliable and representative baseline for comparison with the future (forecast) level of performance is essential to the accurate reporting of benefits for the business case. The question is when should this baseline be measured and established.

If the program is long and there is a lengthy period of time between program commencement and the realization of benefits, the results may be influenced by other factors in the intervening period. Establishing the baseline at the commencement of the program provides data which represents the situation at the time of making the investment decision.

Establishing the baseline immediately prior to the transition may provide an unrepresentative level of performance. At this point, many, if not all, of the stakeholders will be aware of the program and its intentions, and the level of performance may be influenced by the interest shown in this area. The so-called Hawthorne effect[*] suggests that productivity gains may result from the motivational effect on factory workers due to the interest that management (or researchers) take in their work. The results of the Hawthorne experiments, which were conducted in the 1920s and 1930s, have been reanalyzed many times

---

[*] Mayo (1949) undertook a series of experiments at Western Electric's Hawthorne Works complex which concluded that the productivity of the individuals and teams improved when they were aware that they were being observed, thus making it difficult to establish a true baseline level of performance. The accuracy of the Hawthorne results and conclusions has been the subject of debate since the work of Landsberger (1958). Some researchers argue about the data and the repeatability of the experiments, while others attribute the effect to other, often external, factors.

Coordinate and Realize the Benefits 189

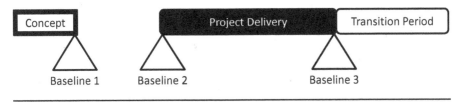

Figure 9.2 When to Set a Baseline

since. However, the fact that the workers knew that they were being observed, and that their superiors were taking an active interest in the results, is credited with influencing the performance of the team. In short, the act of observing the environment may result in changing the actions under observation.

In the context of a program, widespread knowledge that the program has commenced with a specific purpose may cause the performance level to change between the start of one of the projects (at Baseline 2 in Figure 9.2) and the transition to the new method of working (at Baseline 3 in Figure 9.2). If the level of performance has increased in that period, then the baseline will be higher than when it was assessed when making the investment decision. Using this baseline may underreport the total change the program has achieved.

Should both "baselines" be established and used for comparison? As an academic exercise, this would provide an interesting level of data and an insight into the timing of changes and how the program really affected the operational environment. However, establishing a baseline requires a commitment of time and resources, and this may not be a desirable use of the team or budget. In reality, the senior stakeholders, in the benefits management strategy, dictate what the baseline should represent. In many cases, the baseline which will be reported against is based on the level of performance which existed at the time of making the investment decision.

Of course, there are exceptions. For example, driving in Sweden used to be on the left-hand side of the road. In 1963, the decision was made to change to driving on the right, in alignment with other mainland European countries. The switch was made on September 3, 1967, meaning that there was a lengthy period of preparation and implementation of a program to make all of the necessary changes. On the day following the change, there was a noticeable and measured reduction in the number of traffic accidents, compared against the baseline which covered the period immediately prior to the switch. Because of the length of time between the decision to make the change and its implementation, using a baseline closer to the change was reasonable. Insurance claims for traffic incidents fell significantly post-transition. Interestingly, after several months, the number of incidents began to climb again, reaching the baseline levels within two years of the change.

9.2 Transition

The transition period is the interface between the projects and the operational environment. It denotes the point of handover of ownership from the program and project managers to the functional manager and the operational team. On one side of the "fence," the project team is busily completing the work within the scope of the project and delivering the final products before handing them to the other side of the fence—the operational teams. If this period is not carefully planned and managed, it is possible for essential tasks to be overlooked because the teams on each side of the fence expect the other side to undertake the work.

Testing, training, and the delivery of guides, manuals, and as-built drawings could arguably be undertaken by either team. Although this workload is likely to be allocated through the contract, if time and resources are short, the responsibility for delivering these products could be transferred to the other party. Pragmatically, the work must be undertaken to ensure that the transition and post-transition activities can be managed effectively. Bearing that in mind, if there is a need to make changes in the scope of the work and it can be better accomplished by the other party, consideration should be given to adjusting the scope and responsibilities to ensure that the work can be accomplished efficiently. From experience, and anecdotally, what often occurs is that work is switched from the contractor to the client but the work may not be completed adequately. As-built drawings are a good example: At the end of a construction contract, the contractor is often required to hand over a set of drawings detailing the structure as it was actually constructed—as opposed to what was initially designed. In an ideal world there would be no difference, but often decisions, changes, and errors are made on site which mean that the final product differs from the plans and designs. It is important for maintenance and future associated work that the actual state of the final product is known explicitly—just ask anyone who has excavated in a clear area only to break an undocumented pressurized water line, sewage pipeline, electrical or telephone cable. Generally, the contractor is responsible for this work, and there is usually a sum in the contract for its completion. However, the client gains great value from these completed drawings, and it is more important that the work is completed and available for use later. Moving this work out of the contract and having the client complete this is not a problem—the contract payments can be adjusted to accommodate the change. However, risks and liabilities will also be transferred to the client, and this may not be acceptable or desirable.

For a less contentious issue, although still an important one, there is perhaps more flexibility. Instruction and maintenance guides for plant and equipment will be required by the operational team. This is something which may be more easily transferred from contractor to client. The teams who will be using the facility will need this information, and taking responsibility for this work may help the operating teams develop induction and training materials and sessions which are more effective because of the detailed knowledge gained during the development process. In reality, it does not matter who undertakes the work (contractual and risk-related issues can be addressed, or may dictate one preferred approach, depending on the circumstances), but it is crucial that the work is completed to a good standard. One possible outcome is that the work is transferred from one party to another but the work is put off because there is no immediate need for the products—as-built drawings are a good example, where the deliverable is not required until future work is planned, which may be some significant time in the future. As there is little urgency this work may be put off, and given a low priority and finally forgotten. Later work is then planned with limited or no information, which creates a greater need for investigative work or leads to assumptions which may be incorrect.

Transition activities include all of those activities required to accept the new deliverables from a project and integrating the products and changes into the operations as the new form of business as usual (BAU). This may include testing and commissioning, if these are not planned and resourced within the project. Training and induction activities may be undertaken during the transition period.

This phase is often an intense period whose activities are coordinated by the BCMs. The result is that the operational team achieves the outcome.

9.2.1 Induction

When making some changes, it will not be possible to completely prepare the stakeholders in advance. There will be information events which advise of the changes and how they will affect the users; however, it may only be possible to fully induct everybody when the change is made. For example, when moving into a newly constructed building, it is possible to prepare people for the changed environment through videos and presentations, but only when the building is completed and occupied will it be possible to undertake inspections and tours to show the locations of fire equipment and emergency exits. No matter how thorough the preparation is, there will be a need for some induction and reinforcement of previous messages during the transition period.

9.2.2 Training

During the transition there may be a need for training to be conducted. Training could also be completed within the projects and may be the responsibility of the project manager. However, if it is not part of the project scope, the BCM needs to plan for these activities and conduct them, normally as part of the pre-transition phase. There are circumstances under which the training is part of the transition, and these include the following.

- Training requires the project to be completed before it can be undertaken, that is, the participants need to use the new plant, equipment, and systems as part of the training. For example, the early adopters of new equipment are unlikely to be able to use simulators or other equipment to gain a full understanding of the new functionality and capability. Consider the Airbus A380 aircraft; Singapore Airlines purchased the first aircraft, and received it in 2007. Given the differences between this new aircraft and previous models, it would have been possible to undergo some training as the aircraft was under construction. However, it was only possible to complete that training and become totally familiar with the A380 once it was delivered.
- Training conducted during the project or pre-transition phases was not effective.
- The training which was originally part of the project scope has been removed from the project to allow the achievement of project and program targets.
- Previous training is obsolete due to significant changes made during the project.
- The gap between training being conducted and the transition period was too great.

Some of these issues or circumstances may only become obvious as the transition progresses. The BCM is in the best position to witness the need for additional training. This may delay the achievement of the outcome and may impact the timing of the realization of benefits.

9.2.3 Reinforcement

In some instances, the BCM may recognize that, as the transition progresses, there is a need to support the users and stakeholders. This support may take various forms, depending on the enthusiasm and ease with which the users are taking to the new deliverables and systems which are being introduced through the transition. It is not uncommon to see the need for one or more of the following types of action to reinforce the reasons and need for the changes and the longer-term objectives:

- Repeat some of the earlier communications, especially those which detail the context of the program.

- Repeat previous training, perhaps in a modified and consolidated format, to ensure that the skills remain current.
- Hold individual or group discussions/workshops to establish how the transition is proceeding and if there are opportunities for change which may make the transition more effective.
- Explain that the transition must be undertaken and provide instructions regarding how the team must change to comply.

If resistance is encountered prior to or during the transition phase, it must be addressed swiftly. Any resistance will become more obvious during the transition phase. There may be legitimate reasons for this behavior, and the BCM, being the member of the program team closest to the operational teams, should be the person who acts initially. In some situations, the transition must proceed, and any obstruction and objections need to be removed or resolved as a matter of urgency.

Undoubtedly, the best approach to managing resistance to change is to engage the stakeholders early, to ensure that by the time the transition phase is entered, an acceptable solution to the concerns of the users has been agreed and deployed.

9.2.4 Outcome

The aim of the transition phase is to deploy the new deliverables from the projects within the operational environment and, quickly, ingrain the new methods of work as the new BAU. The outcome is achieved when the system becomes sustainable with the new methods of working. The role which is in the best position to determine that this point has been reached is the BCM—who should be appointed to the role from a position of operational authority.

With respect to the BCMs, when the outcome has been achieved, their role within the program is no longer required and they can be released from their initiative-related responsibilities and return their complete attention to their operational and functional duties.

This assumes that the transition phase progresses according to the plan. This is not always the case. What happens if the team are not able to reach the outcome? Or, more likely, what if reaching the outcome is not possible within the definitive timeframe set? In many cases, there is no problem with a short delay. However, sometimes the transition period must fit within a very specific and non-negotiable timeframe. For example, deploying software, testing, and commissioning a system in a power station may need to fit within the scheduled operational shutdown for maintenance. Once that period is over, the power station must start generating power again. A significant asset, such as a power generation plant or mine, cannot be offline too long.

During the transition there should be reviews of the progress conducted by the transition team and BCM. If the work is not progressing in a satisfactory manner, a decision will be made to accelerate the work, possibly by working overtime or introducing additional resources. Introducing new resources may not help, according to Brooks's law,* and may be an expensive option. The BCM, in consultation with other key stakeholders, should make the decision about how to respond under such conditions. A drastic action might be to roll back to the existing system while the problems are resolved off-line and the transition is rescheduled for a later date.

If the construction of a project is slower than planned, the intended occupants may need to remain in their current locations for an extended period of time until the building can be completed and commissioned. If the deployment of software is not going to be completed and tested during the operational shutdown of a mining facility, the team will roll back to the existing system. The operational teams will continur to use the older system, and the implementation of the new software will be rescheduled for the next shutdown—which may be several months away.

9.3 Post-Transition

Once the outcome has been achieved, the environment becomes the new BAU or new way of working. As the teams become more familiar with the new environment, including the new systems and the changes, new ideas may be generated for additional changes or modifications to the system, the training, or the operating environment which are necessary for the realization of benefits and longer-term improvements required for sustainability. The BCM should monitor the situation and determine if further actions or adjustments are required.

9.3.1 Stepping Stones

A number of steps often need to be undertaken once the outcome has been achieved and the new methods of operations have been established. For example, when the transition is completed, there may be a need to recruit new team members with complementary and different skills to the existing team. These

---

* Fred Brooks was the manager of IBM's project to develop the OS/360 operating system. In 1975 he published *The Mythical Man-Month*, which contained the adage that "Adding manpower to a late software project makes it later." This became known as Brooks's law. Although it was derived from a software development environment, the "law" applies to all projects, including construction, infrastructure, and defense.

new skills and expertise may be required before the benefits can be realized in their entirety. The new recruits will need to be inducted into the organization and assimilated into the team. In the post-transition period these tasks are likely to rely on the operational resources to deliver—for example, Human Resources and the team leaders. However, the BCMs must continue their interest in the program and particularly in their responsibilities to ensure that these steps are taken.

9.3.2 Minor Adjustments

After the system goes live and people begin to use it in larger numbers, there may be observations which can lead to improvements in the system. Some of these improvements may not be obvious during the planning and implementation phases of the project, but fresh eyes and different perspectives may highlight them. For example, a new road scheme may have all of the required signage when it is completed and opened for use. However, vehicle drivers may point out that the new round-about system is complex, and additional, and perhaps earlier, signage would allow them to get into the correct lane easier and earlier. This would keep the traffic flow smoother and likely lead to fewer incidents as vehicles try to change lanes at the last moment.

Again, the BCMs are the role most likely to observe the need for minor changes, or have the need brought to their attention by other stakeholders. The BCMs then determine which changes are legitimate and realistic. The funding and resourcing of the changes will depend on a number of organizational factors:

- If the program is still delivering projects, the BCM may be in a position to approach the program manager and make the case for the additional work to be undertaken.
- If the program has been closed, the BCMs may be able to authorize and fund the changes themselves, in their capacity as leaders of the operational teams.
- If the program has been closed, the BCMs may raise the need for a new project to be undertaken with the authorizing body within the organization.

Some of these proposed adjustments may be relatively minor in nature, while others may need more significant investment. In all cases, a business case should be developed to demonstrate the advantages and value from undertaking this additional work. The debate will often revolve around the need for the work in order to realize, or enhance, the benefits within the original program.

Alternatively, there may be an opportunity to develop some emergent benefits by the completion of some additional work.

9.3.3 Reinforcement

As with the transition phase, it may be necessary to reinforce the changes and the need to sustain the new behaviors and methods. During the post-transition phase, users may be tempted back to the older methods, if that is still possible, if they do not see the advantage to the new BAU.

Because the BCMs come from operational leadership positions, it is expected that they will notice that the new behavior is not being sustained and establish the spread and impact of some people, or groups, rolling back to prior norms. The BCMs need to be proactive in addressing this issue, to ensure that support and reinforcement is provided in an appropriate form.

9.3.4 Measuring Benefits

The measurement of benefits should be undertaken by an independent person or group to ensure that the results are fair and unbiased. The BCMs will be involved in the scheduling of activities associated with the measurement of benefits. However, as the BCMs have responsibility for the realization of the benefits, they have an interest in reporting the benefits. Additionally, the BCMs will also be benefit owners and have an active interest in some of the benefits which will be measured.

The activities involved in the measurement of benefits will take different forms depending on the specific benefit being measured. These tasks may be carried out over a period of time to capture all of the data required to establish the size of the benefit accurately. For example, if the benefit to be measured concerns productivity in a workplace, then measuring the productivity on one day is likely to be unrepresentative, and the baseline and each subsequent measurement may require, say, two weeks of data to establish a reliable norm. Similarly, if a new road scheme opens, measuring traffic flows during school holidays or during a holiday period would not present the full picture of the results of the changes.

When considering which benefits will be measured, it is useful to identify those metrics which are currently measured. This may reduce the effort in collecting benefits-related data to a minimum. For example, consider the number of patients presenting to an Accident and Emergency Unit of a hospital. This is a metric which would be gathered in the normal course of the duties of the admissions system. This means that a baseline is easy to establish—due to the wealth and depth of data readily available—and measuring any benefits will automatically be continued in the post-transition period.

There are two crucial elements to be addressed for this activity:

- Ensuring that resources are planned for these activities and that sufficient time is allocated to gather a representative amount of information to reliably compare against the baseline.
- Ensuring that the benefits are measured independently and reported. Too often this activity is omitted to avoid reporting failure to achieve the original goals or because the attention of some stakeholders has moved on to other things.

9.3.5 Decommissioning Obsolete Systems

Where a significant change is undertaken, there is often a period of time when there is an overlap between the old and new systems. This will occur during the transition phase, in most instances, and may under certain circumstances be extended into the post-transition phase. This can be a comfort for the operational team, who will know that, in an emergency, they can roll back to the old system. But it can also be a distraction, with two differing systems live at the same time. This becomes a significant drain on resources with the entry of data into two systems, or maintenance being required to two roads or buildings.

The BCM responsible for the affected area will review progress and the stability of the new system before concluding that the new way of working is embedded in the organization. At this point the BCM will make the decision that the old system is no longer required and can be decommissioned. A risk assessment should be undertaken as part of this decision-making process, because this is a point of no return. If there is a failure of the new system at a future date, it will not be possible to roll back to the old system. If an old road has been decommissioned and is unusable, there is no option to divert traffic along that road in the event that there is subsidence causing the new road to fail with significant potholes or sinkholes. Decommissioning is a major decision which may create anxiety among the stakeholders and should be considered carefully.

The decommissioning itself may be rather simple, as in making an obsolete database inaccessible, or extremely complex, for example, the dismantling and destruction of a power station. The decision that the old system can be decommissioned is essentially an operational decision. As such, the BCM is the role which must make this decision. Resources will need to be deployed to implement the decommissioning, and these may come from the operational teams, and be led by the BCM, or may be undertaken as part of the program, led by the program manager, or implemented as a separate project with a newly appointed project manager.

Table 9.1 records the major activities conducted during each of the phases.

Table 9.1 Activities Conducted in Each of the Three Phases

Pre-Transition	Transition	Post-Transition
Planning transitions	Taking ownership of project deliverables	Monitor the environment
Communicating with users from an early stage	Installation, integration, and commissioning activities	Reinforce new systems
Monitoring projects for changes which will impact the users	Inductions	Measure benefits
Training	Training	Decommission redundant systems
Establishing a baseline for the benefits	Decision to go live/roll back	
Decision that the operations are ready for transition	Achievement of outcome	

9.4 Sustainment

There will also be a need to coordinate the activities in the sustainment plan. Some of these activities will be conducted in the operational environment and will fall within the authority of the BCM. Other activities may need to be

Table 9.2 Summary of Sustainment Activities

Examples of Sustainment Activities	
Conducted by the BCM	Conducted by Others
Modifications to induction activities for new recruits	Development of new recruitment policies
Succession planning for key personnel	Development of forecasts for the new skills needed within the organization
Toolbox meetings (informal meetings of the operational teams to present new information, processes, and develop new skills)	Engagement of new support partners
Reassignment of operational workloads depending on initial performance in the new BAU	Development of long-term, personal development plans for the teams

conducted at a corporate level to ensure that the newly implemented environment can be maintained in the longer term. The significant activities are documented in Table 9.2.

A large number of activities within the project, program, and operational environments must be accomplished before benefits are realized. The coordination of these activities will land primarily on the BCM, although they will be supported by change teams and the program/project managers. There must be a solid working relationship between these actors, and good channels of communication to ensure that there are no surprises or ambiguity.

9.5 A Case Study

A university constructed a new building as a state-of-the-art research and education environment, complete with an administration center for one of its faculties. In addition to several hundred students, almost 1,000 staff members were relocated from existing locations into the new facility. It is easy to focus the attention of the team delivering the project on the physical building—after all, this will account for the vast majority of the total costs. New equipment will be installed, including IT systems, and existing equipment will be relocated. The building is a visible sign of achievement which will be seen for the next 40–50 years.

However, what will happen if the building is completed on time but nobody is ready to relocate? What will happen if the students and other occupants do not like the new facilities? The next 40–50 years will be hard going for them—or, more likely, they are going to look to ply their trade elsewhere.

What are the benefits from a project such as this? There will be a mixture of financial and nonfinancial benefits. Consider:

- Reduction in rental/lease expenses—By consolidating this group in one building, there will be no need for the current accommodations and they can be returned to the owners at the end of the existing rental contract.
- If the existing facilities are owned by the university, they can be repurposed or sold off to realize capital gains.
- New research grants may be awarded. It may be difficult to attribute these benefits solely to the new facility, but there is likely to be some impact associated with the new capability and its unique equipment and features.
- Increased student enrolments—There may be an observable increase in student interest in the new capability, leading to increased student numbers.
- Increased volume and significance of research output—By co-locating the researchers, there may be an increase in creativity and collaboration among groups which were previously separated.

- Consolidation of equipment and technologies in one place may increase use of this equipment.
- New and different events may be attracted to the new facilities—for example, conferences.
- The new administration center may improve the productivity and responsiveness of personnel when dealing with students, leading to increased levels of student satisfaction.
- If the new building is distant from the rest of the campus, students may find it difficult to engage with other students in other locations. There may be a reduction in these interactions and possibly a reduction in student satisfaction—that is, there may be some dis-benefits too.

However, these benefits need to be nurtured and harvested. They will not happen by magic.

In this project the three distinct phases were applied:

- Pre-transition—When the project will be delivered and the teams who will be relocated will be engaged to ensure that their specific needs are considered and addressed as much as possible.
- Transition—Once the building is completed and the people and equipment are moved into the building. This may happen over a period of months.
- Post-transition—The building is occupied and represents an operational environment. Changes may be made as the teams become more familiar with their new home. The new environment should be able to deliver the benefits desired.

9.5.1 Pre-Transition

During the pre-transition phase, the construction project is undertaken. The groups which are to be relocated into the new building are identified, and reference groups are established which should be representative of all of the affected groups, including (in no particular order):

- Faculty members
- Teaching staff
- Researchers
- Security
- Maintenance
- IT
- Student representatives

Some of these stakeholders may be involved in the early design phases of the project. It is important to understand the specific needs of the occupants—consider the impact on benefits of completing the building without taking into account the floor loads applied by some important items of equipment. Costs would rise to address the problem, or the equipment might not be moved into its intended home. Getting the input of the occupants is likely to make the transition smoother. Additionally, the occupants and support services that will manage and use the completed facility will need to know what will be handed over, so they can plan maintenance and operational changes. Security services may have specific and unusual needs imposed by the use of the building and its occupants. Knowing what these requirements are will allow them to be taken into account when designing the building.

In our example, many questions concerning the new environment were raised relating to the need for specialist equipment on the 18th level, the suitability of an open-plan layout, floor occupancy or density, and the need for private rooms and conference facilities.

In some cases, the stakeholders, or their representatives, will be involved in the design process. Beyond this, as more information becomes available regarding floor details, allocation of space, and timing of moves, it will be important to communicate this information widely. It is worth considering spending some time and money to ensure that everyone knows what to expect. Prototype offices can be erected to make it possible to visualize, touch, and experience the new space. Making the transition smooth, and rapid, is likely to lead to the earlier realization of benefits and the avoidance of some dis-benefits.

Workshops were held to engage with the people affected, to help them become familiar with the logistics of the relocation and be aware of any changes to their working environment. This allowed the occupants to relate the design and changes to their own work, and internalize what the changes mean to them as teams and individually. There was a need for training of some of the occupants in the use of new systems and equipment, and for legal requirements such as the appointment of fire wardens and safety officers.

Moving into a new facility will lead to a review of many of the support services such as maintenance, security, and the management of goods delivered. The time to address these issues is while the project is underway and before the transition begins.

A timeline was established and communicated to the affected personnel. This was indicative to begin with and updated as the project progressed.

As more information became available, it was distributed through the reference groups, who cascaded the news to the wider user groups. Changes to the schedule or the design which would affect the occupants were also communicated, to avoid any surprises close to the relocations themselves.

Transition planning was undertaken so that the program team understood where people would be moving from and to which work stations in the new building. It was important to know who was moving, when, and their new location, as well as their network identification, to allow setting up the telephone and data access points so the new occupants could log into the computer system as soon as they arrived.

9.5.2 Transition

Much of this planning continued after the building was complete and the first tranche of people were relocated. This was required because the building was occupied in a number of waves over a period of several months.

Immediately after the building was completed and accepted, the first wave of occupants moved in. The relocation of people into their new environment was followed by inductions performed by the transition team. Presentations and tours were conducted; these were designed to familiarize the occupants with their new surroundings, including emergency responses and exit routes, as well as the locations of services and facilities. These inductions could only be undertaken once the building was complete (although some preparation was completed beforehand). The inductions were most effective once the occupants were in the new building, to keep the content relevant and incorporate any last minute changes to the design.

Some training was required for some of the academics and researchers with respect to new equipment. All of the software used throughout the new building was the same as in other areas of the university, but there were enough differences in the new hardware and building management systems to make short training sessions for individuals and small groups meaningful.

Following the first wave of the planned relocation, some feedback was sought from the new occupants. As a result, the transition team made some changes for subsequent groups, including the preparation of some additional documentation which was left at each workstation to support the initial setting up of PCs and telephones.

The outcome was partially achieved once the first occupants were working in the new building and teaching was being conducted. With each wave of relocation, the outcome was increasingly fulfilled, as the building became more heavily occupied and moved closer to its fully operational state.

9.5.3 Post-Transition

The post-transition period is scheduled for several years, as the required work is undertaken within the previously occupied buildings to allow them to be returned to owners, sold, or repurposed. These decommissioning works will take some time to complete and will be managed as separate projects.

With the new facility fully operational, some changes are likely to be suggested and acted upon to make the day-to-day environment more efficient and comfortable for the occupants.

These activities are summarized in Table 9.3.

Table 9.3 Activities Conducted Within the Coordinate and Realize the Benefits Process

Activity	Purpose	Responsibility	Documentation
Workshops and subsequent planning	Develop and revise the transition plan	BCM	Transition plan
Workshops, meetings, and communication events	Keep stakeholders (particularly users) informed of progress and the purpose of change	BCMs and change teams	
Monitoring progress of projects	Plan for transition	BCM	
Establish appropriate baseline	Create accurate basis for measurement of change	Program manager and BCMs	Updated benefit profiles
Training and induction activities	Prepare the operational teams for the new systems	BCMs and change teams	
Integrate products into operations	Achieve the outcome	BCMs	
Monitor and reinforce the changes	Maintain progress towards the outcomes and benefits	BCMs	
Measure benefits	Confirm progress toward the final benefits	BCMs or program assurance	Revised benefits realization, transition, and sustainment plans Revised business case
Decommission redundant systems	Remove the opportunity to return to the old methods of work	BCM makes the decision—others may become involved	

9.6 Documentation

No additional documentation is created at this point of the benefits life cycle. The program plan, the transition plan, and the benefits realization plan are reviewed and updated as progress is made through the phases.

Ideas for change—whether to any of the plans, the deliverables, or the operational processes and activities—should be addressed in the same way that a project issue would be. This will follow a process allowing the issue to be raised to the role which is authorized to make a decision about it.

The steps in such a process normally are the following:

- Identify and record the issue/idea—Any member of the team or stakeholder community should be able to raise an issue, which may take the form of
  - Request for Information
  - Request for Change
  - Idea for change or improvement
  - Notification of an unforeseen event
- Assess the issue—Before action is taken, an assessment of the issue should be undertaken by an appropriately qualified person. The impact of incorporating the issue into the program should be understood before a decision is made.
- Escalate the issue to the authorized person or group—The person or group delegated to make a decision about the change will depend on the scale of the issue and its impact on the performance of the program or the organization. Generally, the program manager will administer the issue and make a decision if it lies within his or her delegated authority or make a written or face-to-face presentation of the issue complete with recommended actions. Issues should be brought to the attention of the authorized personnel for a decision, who may be the
  - BCM if the issue is an operational matter, or can be managed within the transition plan and the constraints of the benefits assigned to the BCM as the benefit owner.
  - Program manager if the issue can be resolved within the constraints of the program, which means that the funding and resourcing to manage the issue are available to the program manager *and* the type of issue under review has been delegated to the program manager. The sponsor, or higher authority, would be required to make decisions which affect the overall business case.
- Decision—Made to address the issue, along with agreed actions. The action may be to accept a recommendation from the team, modify that recommendation, or do nothing.

- Implementation—The team implements the actions agreed to by the appropriate authority.

Projects and organizations may have documentation which must be completed when addressing issues such as these. This documentation should be used rather than introduce additional templates and documents.

9.7 Summary

The coordination and realization of benefits is a lengthy process which involves the BCMs working within a program capacity and, simultaneously, continuing with their operational duties. The BCMs must manage three phases of activities:

- Pre-transition—When the tasks are associated predominantly with planning and early engagement and communication with the stakeholders. Some of the preparation work required will include training and changing existing methods of operations.
- Transition—During which the technical outputs from the project or program are integrated into the operational environment and they are used by the stakeholders.
- Post-transition—When the benefits are realized and measured, and the new methods of working are sustained.

The process as a whole does not require the full-time commitment of the BCMs, nor their teams. However, it is anticipated that there will be periods of time when a significant effort will be required, especially in the run-up to and during the transition phase.

Some of the issues that should be addressed during this process include:

- Has a suitable and representative baseline of the performance been established?
- Have the pre-transition activities fully engaged the stakeholders?
- Are mechanisms in place for testing and evaluating the effectiveness of the engagement?
- Has the transition phase been planned adequately?
- Have sufficient resources been committed to the transition phase?
- Have the key decision points been identified?
- Readiness for transition
- Go live or roll back
- Achievement of outcome
- Ready to decommission existing systems

- Have resources been allocated to measuring the benefits?
- Have benefits been independently measured?

The next chapter discusses the review of the initiative as a whole and the opportunities to make decisions regarding its future direction.

Exercises and Activities

1. Consider initiatives within your organization. Discuss the initiative and the establishment of baselines for comparison with future benefits.
2. Discuss an example of reinforcement activities and the circumstances in which it was recognized that these were necessary.
3. Provide an example of "decommissioning obsolete systems" and discuss how it was managed.
4. Discuss the role of the BCMs during this process and how they can influence the achievement of the overall objectives.

Chapter 10
Review the Initiative

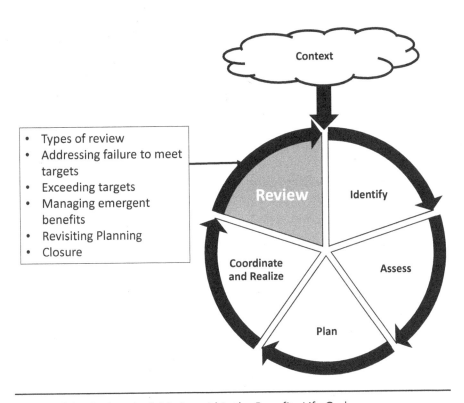

Figure 10.1   Review the Initiative within the Benefits Life Cycle

> *"Examine what is said and not who speaks."*
> – African proverb

> *"Criticism, like rain, should be gentle enough to nourish a man's growth without destroying his roots."*
> – Frank A. Clark

> *"I think it's very important to have a feedback loop, where you're constantly thinking about what you've done and how you could be doing it better."*
> – Elon Musk

Feedback is an important aspect of any system which, if conducted in a timely and constructive manner, will lead to learning, improvement, and better decision making, as shown in Figure 10.1, where this process is used as a mechanism for making decisions. With the generally long timeframes involved in a program, especially when including the realization of benefits, it may be easy for other distractions and priorities to demand the stakeholders' attention and focus. It is essential that the stakeholders remain engaged for the complete cycle of delivery of the initiative, including the realization of benefits.

Additionally, a minor deviation from the target early in the program may result in missing longer-term targets significantly. Monitoring the progress of the early work, adjusting future targets, and resetting expectations of the stakeholders are important facets of the program manager's role. This enables real-time engagement of the people with a vested interest in the investment and allows them to express their ongoing commitment or concern for that investment. Casting this in a more proactive—and positive—light, the program manager has an opportunity to respond to any shortfalls in the forecast benefits earlier in the program life cycle, and to redress problems to ensure that resources are deployed early enough to make a difference and achieve the goals set.

Of course, if early indications are that the program will be more successful than initially planned, the program manager will be in the enviable position of being able to announce that the total benefits will be greater than forecast or will be realized earlier than forecast—or both. In a situation such as this, the stakeholders may make the decision to invest additional resources to take advantage of any opportunities to increase the total value of the benefits.

The whole program should be reviewed at specific points in the life cycle, such as the end of stages or phases, to provide feedback on the progress and performance of the projects and program. The information from these reviews will enable more accurate planning in the future; lessons to be captured, learned, and applied; the management of risks and issues; and engagement with the stakeholders. In a similar manner, reviews of benefits at preplanned points and, if necessary, at regular intervals will allow changes in the work to be undertaken to realize the benefits to be better planned and adjusted to the advantage of the program.

The review of benefits has been mentioned as an activity within the post-transition phase. The process, "Review the Initiative," will be discussed separately because a number of forms of review should be covered and will lead to actions being taken and decisions required by the program team, sponsorship roles, and investors. This process and this chapter will discuss the need for more holistic reviews which address the initiative's ongoing progress and success. The primary purpose of this process is to review the initiative, make decisions regarding its continued progress, and approve changes, if necessary.

A number of reviews may be conducted during a program of change, including:

- Project reviews—Focusing on the progress within an individual project
- Program reviews—Often combining an assessment and evaluation of the progress of the live projects and the outcomes and benefits achieved at a point in the program life cycle
- Benefit reviews—Where benefits are measured
- Business case reviews—Periodic reviews of the business case as a whole

In all cases, it is important that decisions are based on reliable, accurate, and unbiased information. For that reason, reviews should be undertaken by an independent person or groups who have no interest in the results of the review or the program itself.

Table 10.1 provides a summary of these reviews.

In some cases, these reviews may be combined, or at least conducted at the same time. For example, at the end of a phase within a program, there will be a need to determine how the program is progressing compared to the program plan. If the next phase is to be authorized, as planned, or with modifications to the program plan, it is necessary to establish what benefits have been achieved and whether that is aligned to the forecasts within the business case. A Program Review will be required, and a benefits review (unless one was recently completed), because the results from the benefits review will inform the business case review. At this point, three reviews will be conducted, with some overlap and connectivity among them. Coordinating, or combining, these reviews may lead to a quicker and more effective assessment and evaluation of the program's status.

All actions within the "Review the Initiative" process will lead the team to determine which of the other parts of the benefits life cycle should be conducted next. The next tasks following a review may be in other processes:

- *Coordinate and realize the benefits,* if everything is proceeding well and there is no need for any corrective or avoiding actions

Table 10.1 Summary of the Reviews

Review Type	When They Could Be Held	Who Could Be Involved	Purpose of the Review
Project	• At the end of project phases or stages or by exception, when a significant issue has arisen or change is recommended.	Program manager Program management office Independent consultant or expert	• Obtain a good understanding of the status of the project prior to committing resources and funding to the next phase/stage. • To enable decisions regarding future investment and change to be made with full access to reliable information.
Program	• At the end of phases or tranches in the program. Immediately before the next round of investment is required to commit to a number of projects.	Sponsor Program manager Program management office Independent consultant or expert	• Obtain a good understanding of the status of the program prior to committing resources and funding to the next phase/tranche. • To enable decisions regarding future investment and change to be made with full access to reliable information.
Benefits	• At scheduled points in the benefit/program life cycle where specified benefits will be measurable.	Sponsor Program manager BCMs and benefit owners Independent consultant or expert	• Generally applied to measure one or more specific benefits at points identified in the benefits realization plan
Business case	• At milestones in the benefits and program life-cycles. • At regular intervals to enable reporting to higher authorities within the portfolio.	Sponsor Program manager BCMs and benefit owners Independent consultant or expert	• Undertaken to compare the status of the delivered business case against projections at specific points in time. These points are selected to coincide with investment decisions or at times to enable reporting to the portfolio.
Closure	• At a time determined within the benefits management strategy. • Arising out of information presented which suggests that there is little advantage to continuing to measure and review the benefits.	Sponsor Program manager Independent consultant or expert	• To determine if further reviews are required. Have conditions been met which indicate that there review of benefits should cease?

- *Identify the benefits,* if emergent benefits have been identified
- *Assess the benefits,* if the information available indicates that the value of the benefits to be realized will differ from previous forecasts
- *Coordinate and realize the benefits,* if the review has uncovered a need for additional work to ensure that the benefits are optimized

Reviewing the initiative is so important because it keeps the benefits life cycle turning over from one process to the next. Its purpose is to gather the most current information, evaluate the situation, and then make a decision regarding future actions.

Some of the questions related to these reviews, which will be discussed in the following pages, are (in summary)

- Has the program failed to deliver one or more benefits?
- Is the shortfall in benefits (compared to forecasts) simply a timing issue? Has the review been held too early?
- Can the shortfall in benefits be remedied by taking some action?
- Is the benefit shortfall a result of an unsuccessful transition?
- Does the excess of benefits (compared to forecasts) reflect good work, or luck?
- Are opportunities available to improve on the current forecasts and exceed the expected benefits which will be realized?
- How can emergent benefits be managed?
- Has the program reached its ultimate goal?
- Are the benefits sustainable?
- Are the benefits continuing to increase over time, or have they reached a plateau?
- Has the program reached a point at which no further reviews are required, or warranted?

10.1 Addressing Failure to Meet Benefits Targets

Sometimes, progress will not be made according to the plan—any of the plans. As benefits are measured, they may be less than forecast at that point. The business change managers (BCMs) should be proactive in light of this information and engage with the sponsor and other key stakeholders about developing a response to the situation.

Of course, there are a number of reasons for underachievement of results, and the appropriate actions will depend on the cause of the failure to meet targets and goals. For example:

- The failure may be a total absence of one or more benefits.

  It is unusual for a well-designed program to be completed and for one, or more, legitimate benefit to be totally undelivered. It is far more common for the benefit to be realized but to a lesser value than forecast. However, external factors may have surprised the team and changed the environment to such an extent that the benefit has not been realized.

  Remedying this may not be possible. The sponsor and key stakeholders will make the decision to invest additional resources, for example, to implement another transition phase. Since "discretion is the better part of valor,"* a more prudent response may be to stop pursuing that benefit. Resources have already been spent in the so-far-unsuccessful attempt to realize this benefit. A clear, objective decision is required to either deploy more resources, if it is believed that the benefits are still attainable, or to halt actions associated with the realization of the particular benefit.

  This decision will be influenced by, among other things, the magnitude of the forecast benefits, the strategic importance of the benefit, and the visibility of the benefit—meaning how many people will notice if the benefit is not delivered.

- The review may have been conducted too early to capture the full extent of the benefits.

  A delay to a project and/or the transition phase may cause a delay to the timing of the realization of a benefit. A review which is undertaken at a predetermined time (for example, at the end of the fiscal year) may underreport the benefits because they have been delayed and the scheduled review was held too early to confirm their realization or was premature, so that the full value of the benefits was not achieved at that point in time.

  Patience may be the best course of action: waiting until the impact of the delay is known. There may be no cost-effective action which would accelerate the realization of benefits to "catch up."

  The benefits realization plan may be adjusted to show the impact of the delay, and this should be communicated to the stakeholders to allow them to adjust their expectations.

- Similarly, the original estimates of the lag between the completion of the transition phase and the realization of the benefits may have been

---

* "Discretion is the better part of valor" is adapted from William Shakespeare's *Henry IV, Part 1* (ca. 1597), in which Falstaff states that "The better part of valor, is discretion." It is a synonym of "look before you leap," which was first recorded in 1477, and is interpreted to mean that it is better to be prudent than merely brave.

optimistic. There may be evidence that the benefits will be realized, but they have not been realized at the time of the review because they are taking longer than expected, and forecast, to materialize.

Again, patience is one approach, which should be accompanied by advising the stakeholders so they can adjust their expectations. However, there may also be other options to accelerate the realization of these benefits. Because the underreporting is caused by slower-than-anticipated realization, there may be potential to get closer to the estimated level of performance.

- The review may identify the need for remedial actions.

    The review may highlight the need to respond to the shortfall by allocating resources to take action. This action could be required at the project level, the program level, or within the benefits life cycle.

- Did an unsuccessful transition lead to the benefit shortfall?

    The benefit shortfall may have resulted from a poor transition phase. If this is the case, and it has not been addressed previously, the review may lead to the planning and implementation of additional transition activities. These may include further training or inductions, or the modification of operational processes to make the users more comfortable with the changes.

Two effects should be accounted for when planning the transition and the realization of benefits. These may be useful in explaining the results of reviews.

1. Initial Reduction in Performance

    Once a change is introduced, there may be a period during which the performance is negatively impacted. This may manifest by a reduction in productivity, an increase in operating costs, or an increase in the number of complaints. As the new system is introduced, especially when there is a period of overlap with the previous system, resources are stretched to transition to the new approach while maintaining the old methods. There may be a short period when double the amount of work is required and the net benefits may be lower than expected due to the increased effort. With regard to the nonfinancial benefits, the initial level of service provided to customers may be affected because the team is stretched and there may be a lower-than-normal number of people available to connect directly with the customers.

    This may be captured through the costs associated with transition, or by identifying dis-benefits, but the overall result is the same—an initial reduction in net performance which lowers the value of the benefits or delays their realization and measurement.

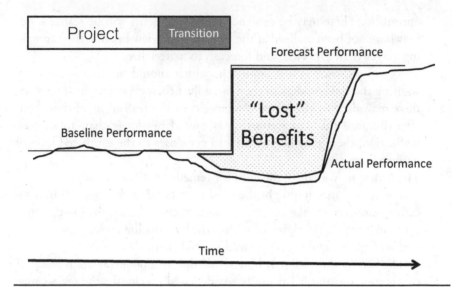

Figure 10.2 Timeline Showing the Impact of Reduced Performance Following the Completion of the Project

Figure 10.2 shows a baseline, existing level of performance which is maintained throughout the project and steps up to the higher forecast level following the completion of the transition period. The wavy line shows the actual level of performance: It is close to the baseline level during the delivery of the project, before a period of lower performance is recorded during, and beyond, the transition period. Actual performance does follow eventually, but may not necessarily reflect the predicted step. This results in a period when the benefits are not realized in accordance with the predictions. This will lead to the measured benefits being less than forecast, possibly by a significant amount.

2. Ramping of the Benefits

Some benefits will be realized immediately the transition period is completed. Others will be achieved over a period of time, with the value of the benefits increasing gradually over a period of time. This ramping up of the benefits to their expected and sustainable value may reduce the measured benefits during early reviews. Hopefully this effect will be identified, and predicted, during the planning processes, but often the impact is only identified in transition and exposed during reviews. In some instances the effect is more pronounced than initially expected, so that even if the ramp is predicted, the ramp is shallower than expected—taking longer for the benefits to reach their final value.

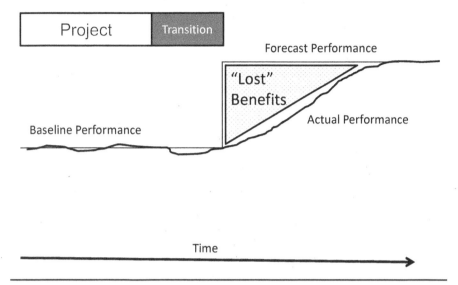

Figure 10.3 The Ramping of Benefits

Figure 10.3 shows a baseline, existing level of performance which is maintained throughout the project and steps up to the higher forecast level following the completion of the transition period. The wavy line shows the actual level of performance: It is close to the baseline level during the delivery of the project and the transition period. Actual performance does show an upward trend following the completion of the transition period. However, the change in performance is gradual, ramping upward toward the forecast improvement. This results in a period when the benefits are not realized in accordance with the predictions. This will lead to the measured benefits being less than forecast, possibly by a significant amount.

3. Exceeding Benefit Targets

The review may find that the value of benefits achieved at the time of conducting the review exceeds the forecast. This is an enviable position to be in, but one which requires some consideration to avoid raising the expectations of stakeholders.

A greater understanding of these results needs to be obtained before a response can be formulated, for example:

- Was the excess of benefits a function of the planned work within the program?

    Original estimates of benefits may have been pessimistic. A cautious approach to planning and forecasting may have been taken in line with an "underpromise/overdeliver" mentality. Exceeding these forecasts will indicate that the program is being successful, but stakeholders must be advised that the primary reason is that the original estimates were deliberately pessimistic to account for some risks.

In other instances, the results may arise from a particularly successful change and transition, and forecasts may be adjusted for the remainder of the program.

- Is this result just plain lucky?

  Sometimes, when the benefits exceed forecasts, it is a result of external factors. For example, if the expected benefit is an increase in export sales, this could be affected by a drop in currency. If the value of the U.S. dollar is lowered in overseas markets, then products and services will become more attractive in those overseas economies as the price in local currency drops and becomes more competitive. Increased sales in these markets cannot be attributed solely to the program, because some of the benefits were realized solely due to external circumstances beyond the control of the organization and program.

  The expectations of the stakeholders may be raised by this result, and it will be important to curb their enthusiasm and advise that future benefits may not be maintained at this level, especially if the strength of the dollar changes again.

- Will the benefits continue at their accelerated rate?

  The early review of some benefits may indicate that they are being realized at rates more quickly than forecast. While there may be good reasons for this accelerated return, such as early excitement or commitment of the team to changes, this may taper off to a lesser level after a period of time. While this is a good result for the business case, it may also set expectations among the stakeholders which are not sustainable. The review must be used to explain this situation to the stakeholders and ensure that they are aware of, and know the reasons for, the amended and more realistic forecasts.

- Are these results sustainable?

  The reviews will offer an opportunity to engage with some of the key stakeholders and bring them fully up to date with the success, or otherwise, of the program. When the review highlights that the benefits realized are more than had been forecast, stakeholders may raise their expectations for future reviews and results. However, the early results may not be representative of the future—that is, the rate of realization of benefits may not be sustainable. It is important that this fact is identified at the reviews and conveyed to the stakeholders to set expectations and avoid confusion and disappointment later in the benefits life cycle.

10.2 Managing Emergent Benefits

As the project and change initiative progresses, and the stakeholders gain a greater understanding of the work activities and the implications in the operating environment, it is likely that additional benefits will be identified which were not apparent during the earlier phases of the work and particularly during the initial definition and planning phases. These emergent benefits can be extremely valuable in supporting the viability of the project.

In many instances, the recognition of an emergent benefit provides the opportunity to review benefits which were previously dismissed or overlooked. This is an exciting opportunity to demonstrate added value to the sponsors and stakeholders. A change to the program plan or the benefits realization plan may be required, and a business case for the additional work for extra reward will need to be presented to the sponsor and other stakeholders. The management of emergent benefits may require additional resourcing and therefore authorization by the investment group.

The major drawback with emergent benefits is the lack of a baseline measure of performance in most cases. The late recognition of these benefits means that the difference between the previous and new performance levels will need to be shown retrospectively. Baselines may be constructed by several means.

- Reference to historical data—if it is available and relevant. For example, in one organization, the implementation of a new financial management system and supporting processes led to the realization that with the increased efficiency of the new environment, there was no need to replace one member of the finance team when she retired. This was not the initial intention of the program of change but was a welcome result. Historical data, from payroll and salary information, meant that the previous performance levels of the team could be reviewed over a period of years to establish a robust baseline. The historical costs associated with that team could be identified and compared to the costs of the smaller team. This information is not always available even in hindsight, however, especially if the benefits are not financial in nature.
- Assumptions based on the experiences of the key stakeholders, for example, the business change managers
- The use of conservative estimates which are defendable during audit. This may lead to a (significant) underreporting of the value of the change, potentially rendering the change as unviable, because the difference between the new level of performance and the conservative estimates used as a comparison is underreported. This approach may be applied when there is a shared resource, and the allocation of time and cost is not clear

between the different groups using the resource. In establishing the baseline, making broad generalizations may be sufficient to provide a robust and defendable starting point.

Emergent benefits occur often in programs and change initiatives, and they need to be recognized and accepted. The key to their effective management is to ensure that the benefits claimed are legitimate—that is, they are attributable to the work being undertaken and not merely the result of fortunate timing and happenstance. When emergent benefits have been identified, there is a danger that the team will enthusiastically attempt to justify and claim them to bolster the business case. These benefits should be rigorously vetted to ensure their legitimacy before they are added to the benefits register.

Once it is confirmed that the emergent benefits are legitimate, their baseline metrics must be established and agreed by the stakeholders to avoid any disagreement in the later stages of the project.

10.3 Revisiting Planning

The planning undertaken for the benefits life cycle should be considered in a similar way to all projects, where the plan is reviewed on a regular basis and amendments made to the schedule to take into account the progress made to date and the changes identified as necessary to improve performance. During the review part of the benefits life cycle, actual progress will be compared to the plan, and there will, at some point, be a need to amend the plan taking into account the success in achieving outcomes and benefits to this point in the project life cycle and the identification of emergent benefits and changing priorities. During this re-planning, the forecasts and predictions for future benefits may be adjusted.

Although the processes within the benefits life cycle are shown as a cycle; this is not always the case. The program manager should consider moving from the current process to any of the others if there is evidence of change from the existing plans and new information is available. The identification of an emergent benefit is a good example; during the process "Coordinate the Delivery of Benefits," an emergent benefit may be identified. At this point the benefit needs to be analyzed to understand its relevance and significance to the program and its business case. The program manager should "move" at least mentally into "Identify Benefits" and work through the processes to develop a benefit profile, integrating the new benefit into the plans and the business case by working through "Assess Benefits and Plan Benefits Realization." There is no single pathway following the processes from the start to finish of the program. It is likely that events and risks will arise which require moving to another process.

10.4 Closure

At some point the initiative must be declared complete. Either all of the benefits expected have been realized, or the benefits have reached a value above which they will not increase anymore and there is no advantage to continuing with the post-transition actions or the measurement of the benefits. The review will provide the information required to make the decision that the initiative should be closed by answering one or more of the following questions.

- Has the full quota of benefits been realized? Once the overall objective of a change program has been accomplished, the program should be closed. With respect to the management of benefits, the same is true. There should be a point at which a decision is made to close the initiative and cease the work required to realize and measure the benefits. The sponsor should be the person to make this decision, and the conditions and timing should be recorded in the benefits management strategy.
- Are the benefits sustainable? If the original goals have been met but the benefits are not sustainable, the initiative should not close without ensuring that actions are taken within the normal environment to manage the benefits over the longer term.

   Consider the case of Sweden changing the direction of traffic (from driving on the left-hand side of the road to the right). Initially, there was a significant reduction in the number of traffic accidents and insurance claims. However, over time these benefits were not sustained—this might be attributable to drivers taking great care immediately following the switch before becoming somewhat complacent and reverting to their previous driving styles and attitudes as they became more familiar with the changes. Without further actions, the benefits were not sustainable in the longer term. It should be noted that this initiative was not designed specifically as a means to safer driving, but with hindsight may have been an opportunity lost.
- Will the benefits continue to increase, or have they reached a plateau?

   Benefits may increase initially to a particular level and then rise no further. The leveling-off may have been anticipated, or may be a temporary phenomenon, or it may be possible to take action to reinvigorate the benefits and increase the growth of benefits. In addition to measuring the value of the benefits in total, at a point in time, the reviews should be more proactive and inform the decision making within the program, investment, and organization.

   Based on the findings and recommendations of a review, further actions may be planned, which can then be implemented. Alternatively,

there may be indicators that the benefits will continue to increase, or are at their upper limits.
- Is there value in continuing to monitor and review the benefits?

    Closely linked to the above point, there is a temptation, especially with a successful initiative, to continue to review and report the situation regarding benefits. A successful initiative will generate good outcomes and benefits in line with, or exceeding, the forecasts. There will come a point where one or more of the following conditions will be met:
    - The initiative reaches the point in the benefits life cycle where the benefits management strategy determines that benefits will no longer be attributable to the efforts of the program.
    - The time between the delivery of projects and the review is long enough that the benefits which will be realized in the future could be influenced by external factors. In this case, it will be difficult to claim that the benefits were the results of the efforts invested, since they may have occurred due to actions and factors outside the control of the program.
    - Some of these factors may be other projects, programs, and initiatives being conducted within the performing organization.
- Have all of the anticipated benefits been realized? Whether the benefits are realized as per the original schedule, or not, is not necessarily an issue of importance. However, once the targets are met, there may be little value in continuing to measure and review the benefits in the future. The

Table 10.2 Activities Conducted within the Review the Initiative Process

Activity	Purpose	Responsibility	Documentation
Review progress of the initiative	Assess the progress and success (to date) of the initiative	Sponsor, program manager, BCMs, and program assurance	Review report
Workshops and subsequent meetings	Present options and make decisions regarding future of Initiative	Sponsor, program manager, and BCMs	
Workshops, meetings, and use of reviews to assess the progress of the initiative	Determining the next steps in the benefits life cycle	Program manager and sponsor	Updated documents and plans
Review the final benefits	Determine whether the initiative should close	Sponsor and sponsoring group	Benefits closure report

information gathered to that review point may allow the development of an accurate model for the growth and prediction of future benefits. When it is clear from the records maintained to this point that the original objectives have been met and there is a reliable method for calculating future benefits, there is little value to be gained from dedicating resources to reviews. An accurate assessment of the results and review of the final business case can be made at this point.

There may be evidence that the benefits have reached a peak above which they are unlikely to rise. This evidence may be based on successive reviews showing the same total benefits, suggesting that continued action may not deliver more benefits.

When one or more of these conditions is met, the sponsor, in consultation with other key stakeholders, may decide to close the benefits processes and initiative. Table 10.2 summarizes the activities which generally take place during the process.

10.5 Documentation

Although the reviews may lead to updating other documentation, only two new documents need to be considered for the benefits review process itself:

- Review report
- Benefits closure report

10.5.1 Review Report

The review report should be a brief summary of the review results together with recommendations for actions. The purpose of the report is to enable the sponsor and other investors to make decisions regarding the continuation of the program. The review report should include:

- Date of review.
- Participants in the review.
- Timing and purpose of the review—with respect to the benefits life cycle. This will enable the recipients to put the report into context.
- Measured value of benefits—depending on the individual benefits, this may be a single consolidated financial sum, or a list of the benefits and their separate values

- Planned value of benefits at this point—the anticipated value of the benefits to the time of the review
- Comparison of actual versus planned benefits—a discussion of the differences between the planned and actual results
- Anticipated results—a discussion and explanation of deviations from the benefits realization plan due to performance to date
- Business case review—the status of the business case should be reported, with explanation of current or projected shortfalls or overruns
- Recommendations—the program manager presents the facts and explains the situation at the review *and* recommends future actions based on performance to date.
- Acceptance—the report should be signed off by the sponsors, indicating their acceptance of the report and the decisions (if applicable) made regarding follow-up actions and the recommendations.

10.5.2 Benefits Closure Report

The benefits closure report is similar to the review report with changes due to the final nature of the report:

- Date of review.
- Participants in the review.
- Measured value of benefits—Depending on the individual benefits, this may be a single consolidated financial sum, or a list of the benefits and their separate values.
- Attributable future benefits—The value and substantiation of future benefits which may be claimed by this initiative, with an explanation of the reasons for claiming these benefits.
- Planned value of benefits—The anticipated value of the benefits to the end of the benefits life cycle.
- Comparison of actual versus planned benefits—A discussion of the differences between the planned and actual results.
- Business case review—The final status of the business case, with explanations of shortfalls or overruns.
- Lessons learned—A summary of lessons learned should be presented for consideration and dissemination.
- Acceptance—The report should be signed off by the Sponsors, indicating their acceptance of the report and the decisions (if applicable) made regarding follow-up actions and the recommendations.

10.6 Summary

The review of benefits overlaps with the coordination and realization of benefits processes and is spread over a lengthy period of time. However, the workload is sporadic, but intense. Reviews should be scheduled following the completion of a transition phase and closed when the sponsors confirm that the final conditions have been met. The reviews undertaken during this process should be considered as holistic assessments of the complete initiative at a point in time rather than only the measurement of one benefit.

The program manager administers the reviews, but consideration should be given to including independent reviewers to ensure that there is no bias in the reporting and that, particularly, projected shortfalls in the value of benefits are raised at the earliest opportunity and can then be addressed by the appropriate authority.

Some of the issues addressed during this process should include the following:

- Is there independence in the review and reporting of the status of the initiative?
- Are there actions which should be implemented to improve the status?
- Based on the findings of a review, what should be the next step in the benefits life cycle?
- Have any of the conditions for halting benefits management been reached?
- Achievement of targets.
- Sustainable benefits.
- Benefits have reached a plateau.
- Benefits are no longer increasing.
- The benefits realization strategy has determined that activities should stop.
- Has the initiative reached a point at which the benefits measurement and reporting should be stopped?
- Have lessons learned been recorded and acted on?

The next chapter will discuss methods for embedding the mindset for benefits realization management within the initiative and the organization.

Exercises and Activities

1. Discuss how, and under what circumstances, the process of reviewing the initiative would trigger each of the other processes in the benefits life cycle.

2. Figures 10.2 and 10.3 show a situation where benefits are not achieved (identified as the "lost benefits") as forecast. Provide examples of how this may arise and the options for addressing this situation.
3. Provide an example of an emergent benefit and discuss how it was identified and managed.

Part Three
Embedding the Practices

Chapter 11

Embedding Benefits Realization Management into Organizations

Benefits realization management (BRM) should not be seen as an additional layer of governance and management practices which must be undertaken in addition to other daily chores. BRM must become embedded into the existing activities within the performing organization and be accepted as a value-adding part of the investment decision-making process. Embedding new practices requires two main activities:

- Induction of personnel into the new practices
- Integration of new practices into the environment to the extent that they are embedded within the standard operating procedures

This chapter will present some ways of embedding these practices to avoid the perception that additional work is being thrust on the delivery and operational teams. These experienced-based suggestions have made successful contributions to the introduction and adoption of benefits-related practices.

11.1 Change the Conversation

Benefits and BRM should become part of the everyday language of the organization with respect to all projects, programs, and investments. When initiatives, programs, and projects are conceived, the first questions asked should be

- What benefits are expected?
- Are the investments supported by a valid business case?
- How does this support corporate objectives?

Bringing the focus to the benefits forces the involvement of key stakeholders who are responsible for investment decisions. The introduction of BRM into the management environment ensures that the effort required in following the benefits life cycle is used transparently in the making of decisions. The broader the knowledge of and interest in benefits is, within the organization, the better able people will be to make a contribution in a meaningful manner.

Agreeing on the language to be used is important. There is great value in imposing a consistent set of terminology across the organization and the stakeholders. Although the terminology can be tailored to take into account the technical and management language which may be familiar to all parties, there are significant advantages to the introduction of new terms. These advantages include:

- Visibility—Applying new terms demonstrates that there is a new approach.
- Clarity—The new terms can be defined in a glossary of terms. This will avoid any confusion with existing terms in common usage.
- Focus—This ensures that the discussion and priorities are focused on the new approach.

This new language should be used and reinforced at every opportunity so that it becomes a natural part of the management environment. The language must be changed and initiatives need to be discussed in terms of their opportunity to generate benefits rather than costs to be managed.

11.2 Enforce the Development of Benefit Profiles

Instituting the rigorous activity of writing detailed benefit profiles is one of the best, and quickest, ways to encourage widespread focus on BRM. It is important that this practice is taken seriously to avoid it becoming another bureaucratic requirement. A clear and comprehensive understanding of the benefits expected from an investment will cast a new perspective over the scope of the work and its purpose.

In practice, the introduction and discipline of recording the details of each benefit and obtaining agreement among the key stakeholders has led to an easier and swifter acceptance of BRM as a value-adding practice which supports investment decisions. This one document, or set of documents, provides visibility and clarity among the teams and stakeholders.

Documenting the benefits in depth ensures a common understanding of the objectives of the investment and opens the reporting of progress to include updates on the realization of benefits. This has been found to be one of the quickest approaches to obtaining support for BRM and for gaining support from senior stakeholders.

11.3 Apply Successful Delivery Mechanisms

Proven delivery methods for managing programs and projects should be adopted. There are a number of standards, practices, and methodologies for the delivery of programs and projects. Rather than develop methods in a "bespoke" manner, one of these proven approaches should be adopted, and, if necessary, tailored, within the performing organization. This provides a consistent approach to the initiatives, and, over time, a greater understanding of the roles within the team and how they can make contributions to the successful implementation of the investment.

The selection of a method will be influenced by several factors, and each organization should select a method which can be integrated into its working environment. This will improve the likelihood of success by:

- Creating a consistent approach to the management of investments
- Enhancing visibility of the progress of initiatives
- Making the on-boarding of a new team, and team members, easier, because the environment and practices are established
- Applying methods which have been successful in similar organizations and industries

Using existing techniques which have a track record of some success provides some comfort that the delivery of programs and projects is likely to be completed within their constraints. This is one way in which the development of a benefit realization culture can be undertaken with minimal change to the organization's environment.

11.4 Integrate BRM with Existing Organizational Processes

The practice of managing benefits should not be seen as an additional component of management tasks. BRM, to be effective, must become integrated into the working practices of the organization. In the same way that program and project management approaches need to become part of the day-to-day activities of the management teams, BRM needs to become embedded in the daily tasks.

In practice, this means combining documentation, reporting, and decision making from the regular operational management activities with existing program and project management practices. This will result in the reporting of benefits coinciding with other progress reporting and, potentially, with the quarterly, half-yearly, or annual reporting cycle within the organization. BRM should become part of the organization's DNA.

The integration must be undertaken with sensitivity to the organizational culture and practices. My preferred word in this context is "sympathetically." The integration needs to be undertaken in a manner which is sympathetic to the personnel and environment. Forcing new practices on people and teams is often unsuccessful. It is more effective to introduce the changes in a manner which accepts existing tasks and sympathetically implements value-adding extras which can be efficiently incorporated into the workplace. This requires careful consideration of what will add value and how to introduce new tasks in the least disruptive manner without losing the purpose of the change. As discussed earlier, early wins are important: Ensuring that the changes are visible and are acted on to add value will enhance the effectiveness and acceptance of those changes.

11.5 Induct All Stakeholders

If people (team members and stakeholders) are expected to participate effectively in the management of benefits, they need to be inducted into the "system" so that they understand their role and how they are expected to contribute to the team and the decision-making process. Generally, in programs there are some senior and influential people involved as BCMs and within the sponsoring group. These successful executives will have experienced different approaches to the management of investments and may have preconceived ideas regarding the role they should play in the initiative. To optimize their input and gain value from this valuable resource, it is important that they are inducted into the initiative and the BRM processes. This will allow them to commit time to the initiative when they can contribute most effectively.

Other team members, and particularly external stakeholders, may never have been involved in initiatives designed to realize benefits in a disciplined way. These stakeholders must understand their roles if they are to be able to participate without getting frustrated.

Induction can be affected in several ways, including:

- Workshops—Presenting to a group with the opportunity for a question-and-answer session. These are normally relatively short workshops of one to two hours.
- One-on-one meetings—Small meetings with key stakeholders to guide and support them in their introduction to the initiative.
- Online resources—Webinars and presentations which can be accessed individually by a new team member.
- Handbooks—A handbook or online resource which can be read as an introduction to the initiative or referred to at any point throughout the initiative.

These approaches are most effective when there are follow-ups to the initial induction in the form of mentoring or coaching. These may be scheduled on a regular basis or at specific points in the initiative to prepare the team and stakeholders for the next involvement. They can become valuable opportunities for the team to learn lessons and be guided in their preparation for future involvement. Initiatives can last a significant period of time, making the initial induction important at the commissioning of the program but less effective for later parts of the work. The need to reinforce these inductions arises from the time passing between the induction and the participation in tasks. Additionally, in a lengthy initiative there are likely to be many changes in the participants, which will increase the need for a program of induction and reinforcement activities to be scheduled.

As the organization matures with respect to BRM, there may be less of a need for formal inductions because the participants will be more fluent in the process. However, there will always be new and inexperienced stakeholders who will need the inductions, or they will feel excluded from participating and, as a result, may become resistant to and unsupportive of the changes.

11.6 Establish a Single Sponsoring Group

One option to consider is to establish a sponsoring group for all investments across the organization. This single group would be the sponsoring group, or form the core of the sponsoring group, for all initiatives. This would allow visibility of all initiatives—a portfolio view, if you like—across the organization, enabling the sponsoring group to prioritize resources in a wider context.

A single group will become familiar with BRM and associated activities and is likely to apply them consistently. This approach can strengthen the relationships among the program manager, BCMs, and the sponsor. It will require some of the senior managers to commit time to the portfolio, but it has resulted in an increase in the maturity in a number of organizations.

Creating a single sponsoring group for all initiatives has the following advantages:

- Ability to prioritize and transfer resources between initiatives
- Visibility of the portfolio at a senior level within the organization
- A consistent approach applied across all initiatives
- Ability to supplement the sponsoring group with additional stakeholders
- BRM highlighted as an essential management approach at the highest levels within the organization

11.7 Focus on the Significant Benefits

It is tempting to "claim" as many benefits as possible to add substance and value to the business case. This can also attract more support for an initiative by reaching out to more stakeholders as the range of benefits expands. However, a large number of benefits may become a distraction and a drain on resources. Each benefit needs to have a baseline metric measured and needs to be measured once it has been realized (and possibly at several points in between). This requires some effort, and the magnitude of a benefit needs to exceed this effort significantly. One method to address this issue is to set a minimum limit for the value of a recognized benefit which will contribute to the business case.

While it is difficult to provide universal guidance regarding the minimum quantities of benefits to be measured, some perspective should be offered. Considering financial benefits alone, any benefit which is less than 8% of the total cost of investment is unlikely to be significant. This limit the number of benefits being measured to 12, but in all likelihood there will be fewer. This is not to say that benefits of lesser value are not important; however, the establishment of a baseline and the measurement of partial and full achievement of the benefits is a significant cost for the program. These costs will eat into the value generated without adding a great deal of value to the business case. For smaller investments, claiming a benefit of value less than $5,000 may be difficult to verify economically.

These figures should be considered as general guidance only, and it is noted that other (less financially valuable) benefits may be included in the business case for a host of reasons, such as:

- The benefit may be strategically important.
- The benefit may relate to a specific key performance indicator (KPI) or metric.
- The benefit may be particularly important to a key stakeholder.
- The benefit may be easily measured—for example, a reduction in number days lost to sickness—and may be reported already.
- Nonfinancial benefits may need to be reported.

A better approach is to focus on the significant benefits. Saket (1986) demonstrated that construction projects consisted of a small number of items that accounted for the vast majority of the project costs. These were referred to as the cost-significant items. In fact, it was found that these items followed the Pareto principle,* commonly referred to as the 80/20 rule. The Construction Management Research Unit at the University of Dundee (Scotland) has applied the concept of significance to a wide range of scenarios. Anecdotally, and from experience, this is certainly realistic. It is common to review a business case which contains a large number of benefits. While many of these are real benefits and attributable to the initiative, others can be removed from the business case because they are not attributable or legitimate. The remainder follows a pattern similar to the 80/20 rule.

In one review of a significant ($60 million) IT program, the business case included a total of 56 benefits and 8 dis-benefits. Analysis of this business case allowed for the removal of some benefits which were duplicated and some which were not valid. The remainder were reviewed and considered in depth. The result was that the final business case included 6 benefits and 1 dis-benefit. This significantly clarified the objectives and reduced the costs of management and review with no loss of granularity. The integrity of the business case remained intact because the benefits removed added very little value.

Applying this approach to benefits, identifying the vital few benefits which contribute to the 80% of the total value of benefits recorded within the business case will greatly clarify the direction and focal point for the initiative. This will have a number of advantages for embedding the practices of BRM:

---

* Italian economist Vilfredo Pareto found that 80% of the land in Italy was owned by 20% of the population. The concept that the vital few (20%) account for the majority (80%) of the results has been applied to many situations.

- Reduces the cost of monitoring and measuring benefits
- Simplifies reporting
- Assembles the "right" stakeholder groups based on the focus of the benefits
- Simplifies communication
- Sets expectations differently and in a more manageable way
- Retains stakeholders' interest and, hopefully, participation, because of the focus on a small number of important benefits

Guidance regarding the measurement of benefits should be recorded within the benefits management strategy.

11.8 Substantiate the Attribution of Benefits

The key advice with respect to attribution is to be strict. Do not accept unsubstantiated benefits into the business case! The inclusion of benefits which are not legitimately attributed to the initiative will distract attention and effort from the true purpose of the investment. This will create confusion and frustration among the stakeholders and team. Guidance regarding the attribution of benefits (as well as details for the resolution of any conflicts) should be recorded within the benefits management strategy.

The following scene from the television series *The West Wing*, written by Aaron Sorkin, discusses how correlation does not necessarily mean causation.

President Josiah Bartlet:	27 lawyers in the room. Anybody know "Post hoc, ergo propter hoc"? Josh?
Josh Lyman:	Uh, uh, "post" - after, after hoc, "ergo" - therefore, "After hoc, therefore" something else hoc.
President Josiah Bartlet:	Thank you. Next?
Josh Lyman:	Well, if I had gotten more credit on the 443 . . .
President Josiah Bartlet:	Leo?
Leo McGarry:	After it therefore because of it.
President Josiah Bartlet:	"After it therefore because of it." It means one thing follows the other, therefore it was caused by the other. But it's not always true. In fact, it's hardly ever true.

Just because a benefit is realized following an initiative does not mean that the initiative caused the benefit. Demonstrating the link between the project, program, and initiative and the benefit is crucial in retaining focus on the results and avoiding wasted effort.

There may be many pressures on the delivery team to record additional benefits, but the team should be supported by independent reviews and the sponsoring group to maintain the integrity of the business case.

11.9 Test the Legitimacy of Benefits

It is easy to become distracted with quantitative issues regarding the benefits, for example:

- Is the value of a benefit $150,000 or $175,000?
- Will the benefit be realized six months after completion of the transition, or eight months?

These are important issues and should not be ignored, because an accurate business case document will aid with decision making within the initiative and increase the likelihood of success. Guidance regarding the legitimacy of benefits should be recorded within the benefits management strategy.

However, the following questions are at least as important:

- Is the benefit attributable to the work conducted within the initiative?
- Can the benefits legitimately be claimed because future performance metrics (budgets) have been adjusted?

Organizations which are stricter about claiming benefits and discarding those that cannot be legitimately attributed to the initiative will be more successful because they will truly recognize the need for coordination of initiative and operating performance criteria. As a result of this greater understanding, these organizations will invest in those initiatives which will make a difference to the performance of the organization and the stakeholders.

Introducing a strict review of benefits and selecting only those which are legitimate is one way to simplify and clarify the objectives of the initiative. This will make the implementation of BRM more transparent and will avoid double-claiming benefits and claiming benefits which have not been translated into adjustments in future reporting and forecasting.

11.10 Beware "Double Dipping"

"Double dipping" occurs when two, or more, initiatives, programs, and projects claim the same benefit. This may occur by coincidence due to a lack of communication between two investments, or it may occur deliberately because of a need to hit targets. Regardless of the reason for double dipping, it needs to be remedied if the initiatives are to be reported accurately and with integrity.

Guidance regarding the resolution of double dipping should be recorded within the benefits management strategy. This may be done by instructions about how the benefits should be allocated and the mechanism for addressing any disputes. Having a portfolio management office, or a single sponsoring group, with visibility across all investments will ensure that a single body has an overview of the distribution of benefits and can act if necessary.

11.11 Apply a Model for Change

The model for implementing change put forward by Kotter (1996) can be applied to initiatives and the management of benefits. Many initiatives focus specifically on changing the existing operational environment to one which offers greater advantages.

Figure 11.1 shows an adapted version of the Kotter change model (Kotter, 1996) and relates it to activities within the benefits life cycle. Efforts have been made to map these two models because both focus predominantly on changes. The alignment to the adapted change model is (starting at the beginning, the base of the diagram):

- *Determine the driver for change.* Creating a need for change encourages action. In the case of BRM, this is likely to be the commissioning of an initiative. If there is a strategic or urgent imperative to act, it is more likely that leaders will commit to changes. In terms of BRM, establishing the context of the initiative and identifying its drivers is the process during which the need for change is developed and agreed.
- *Identify the key stakeholders.* The key stakeholders represent interested parties who have the authority to direct the change. This group will form the sponsoring group.
- *Agree the objectives.* Programs often produce a document called the "vision" or "vision statement." This relatively brief document records the long-term objectives of the program. Recording the benefits management strategy and agreeing on the overall objectives, including the business case, will provide a shared vision which is accepted by all of the key stakeholders.
- *Communicate the path to the objectives.* Spreading the word about the initiatives and the need for change is an important part of the BRM processes. The program manager and the BCMs are the primary roles who will undertake this function, although there are occasions when having the sponsor be visible as the figurehead in terms of communication can be

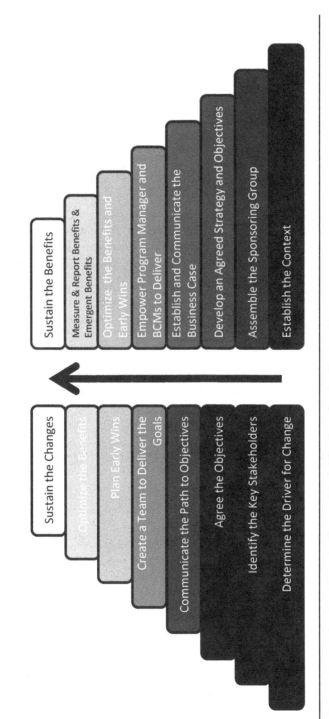

Figure 11.1 Adapting Kotter's Change Model for Benefits Realization Management

valuable. In BRM, the business case is the key document for justifying the initiative and will explain the reasons for the change.
- *Create a team to deliver the goals.* The program manager and the BCMs are empowered to lead and manage different aspects of the initiative. These key actors represent the sponsor on a day-to-day basis and will be directed by the sponsor.
- *Plan early wins.* Early wins are important in BRM to maintain the interest of stakeholders. Regular reviews and reports of the benefits attained will also demonstrate the progress toward the longer-term goals.
- *Optimize the benefits.* The Coordinate and Realize the Benefits and Review the Initiative processes are designed specifically to assess the progress and success of the initiative. Importantly, this opportunity should be taken to be proactive. The BCMs, in particular, will be in a powerful position to consolidate achievements at the point of the review. Further actions can be taken to enhance the emerging and newly identified benefits and build on the initial success of the initiative.
- *Sustain the changes.* The changes implemented by the initiative need to be sustained over the long term so that the benefits continue to be accrued, and there is no danger of regressing to the previous, less-efficient practices. Implementing the sustainment plan enables the organization to institutionalize the change.

11.12 Be SMART

A good approach to defining benefits is to apply the SMART approach. SMART is an acronym for:

- *Specific*—Benefits should be defined in clear and unambiguous terms.
- *Measurable*—Benefits must be measurable and based on an improvement in some performance metric from an agreed baseline.
- *Attainable/Achievable*—Goals should always be set in terms which are achievable. Targets should be set which create a challenge, but forecasting benefits should be based on sound data and judgment and should be free from bias.
- *Relevant*—Research (Dolan, 2002) has shown that goals are more likely to be met when they are agreed by the participants and have meaning, or relevance, to those participants.
- *Time-bound*—Benefits should have time-bound target dates for realization.

Defining benefits rigorously to ensure that there are no misunderstandings among the stakeholders creates a common focal point for the initiative. Reduced ambiguity in the definition of goals and targets will help communications and decision making. The SMART categories will support a disciplined approach to describing benefits and the creation of benefit profiles.

11.13 Engage Stakeholders

Stakeholder engagement is an important element of any project and becomes more significant as the projects become larger and more extensive in their impact. Programs, initiatives, and portfolios require an even greater effort applied to this discipline, as the number and power of the stakeholders increases. Engaging with the stakeholders requires considerable effort from the program manager, BCMs, and sponsor, and in many cases, other members of the sponsoring group.

A strong stakeholder engagement strategy and comprehensive communication plan should be developed within the program, usually by the program manager, and these should be implemented through dedicated resources. The program manager and the BCMs should have strong communication skills and other interpersonal skills:

- Persuasive communication skills
- Leadership
- Team-building skills
- Stakeholder engagement capability with a wide range of stakeholders
- Empathy
- Team-motivating skills

In the absence of these particular skills, they must have people around them with those skillsets and expertise.

The engagement of stakeholders should be given a high priority, with early and proactive engagement preferred to simply sending messages to interested parties. This discipline is an opportunity to assemble Kotter's (1996) guiding coalition of supporters and have its members influence and encourage others who may be more resistant.

Stakeholder engagement needs to be a visible and integral part of any endeavor and initiative. It is never too early to engage with the stakeholders in a complex and impactful change initiative. Change is often an emotive rollercoaster, so getting some (or all) of the key stakeholders onside early and keeping them engaged will reap rewards—especially if there are problems and difficulties further down the line. Stakeholders who have a stronger understanding of the benefits management environment, and the initiative, will be in a better position to help in difficult times—and there *will* be difficult times!

Energy devoted to stakeholder engagement is never wasted—sometimes, it just takes a little time to show visible results.

11.14 Conduct Independent Assurance and Reviews

Assurance must be independent of the program team. However, in many instances assurance functions and reviews are conducted internally. With the best will in the world, that independence can be compromised if there is insufficient separation between the implementation and assurance teams.

Considering the magnitude of investment decisions which are based on assurance activities and reviews, the sponsor and sponsoring group will feel most comfortable knowing that the advice being provided is independent, considered, and can be relied on. Ensuring that there is distance between those implementing the initiative and those reviewing it is imperative.

The use of truly independent people and teams to undertake reviews and assurance activities does incur additional costs. However, given the scale and magnitude of programs and their benefits, the investment in assurance to create reliable and unbiased feedback mechanisms is worth the associated cost. The independence provided in this way will counter strategic biases and avoid the pressure to comply with political decisions and views.

11.15 Create Champions

Closely linked to the importance of engaging stakeholders, the concept of appointing benefits champions will help raise the profile of BRM within the organization and the initiative. Although there may be number of candidates for this role(s), to be effective the champion should have the following attributes:

- Respect from others within the organization/initiative
- A track record of leading or managing successful initiatives
- Credibility within the organization
- Authority to make decisions within the benefits environment
- Strong mentoring and coaching skills

The champion's role is to increase awareness of the value of adopting a more rigorous discipline to the management of benefits. As a figurehead for BRM, the champion can advise and develop benefits-related skills in the delivery teams and operational groups to spread the capability within the organization.

11.16 Summary

BRM must become part of the recognized and supported methods for managing investments—programs and projects. The relevant techniques and processes should be melded with the existing organizational practices. BRM should become an intrinsic part of the investment and change environment.

Some effective approaches to embedding BRM within the organization are

- Change the conversation
- Enforce the development of benefit profiles
- Apply successful delivery mechanisms
- Integrate BRM with existing organizational processes
- Induct all stakeholders
- Establish a single sponsoring group
- Focus on the significant benefits
- Substantiate the attribution of benefits
- Test the legitimacy of benefits
- Beware "double dipping"
- Apply a model for change
- Be SMART
- Engage stakeholders
- Conduct independent assurance and reviews
- Create champions

BRM is conducted in a complex and risky environment. Application of the approaches and processes described in this book will clarify the focus of the investments and initiatives and ensure that the delivery of the benefits is kept front-of-mind for the key stakeholders and those delivering the outputs and changes needed.

The delivery of benefits is not a simplistic, linear process. It is much more complicated and complex. As we described in the Introduction, if there is a malaria outbreak in Borneo, it can be eradicated using chemicals. There may be dis-benefits which affect some of the stakeholders. Lack of understanding of the system which is undergoing the change may result in actions which have an unanticipated and negative consequence. This is what the Dayak people found when their roofs caved in, disease increased, and the rat population increased. This sort of result leads to further expense and sometimes extreme remedial responses—imagine parachuting cats!

Exercises and Activities

1. Discuss how the induction of stakeholders can be accomplished.
2. What would be the most difficult aspects of embedding BRM into your organization?
3. Discuss how the implementation of three of the approaches could make a significant improvement to the adoption of BRM.

Appendix I
Documentation

Benefits Management Strategy

Purpose	Define and record the ground rules for the management of benefits within the initiative	"What" and "how" benefits will be approached in the initiative
Written by . . .	Program manager	Generally with consultation with the sponsoring group
Approved by . . .	Sponsor	
Contents	Definitions and terminology to be used	
	Responsibilities with respect to benefits	
	Baseline requirements	Instructions for the methods and timing of establishing baseline performance metrics
	Requirements for measuring the benefits	
	Overall duration of benefits-related activities	The period over which benefits will be measured

Benefit Profile

Purpose	Clearly defining each benefit	Create a basis for agreement and common understanding
Written by . . .	Business change managers (BCMs)	
Approved by . . .	Sponsor	
Contents	Description of the benefit	
	Projects on which the benefit depends	Delays in the projects will delay the realization of benefits
	Business changes required	
	Timeframe over which the benefit will be realized	
	Method of measurement for the benefit	
	Baseline metric	Timing for its measurement and performance metrics

Benefits Register

Purpose	A single source of key information regarding the benefits	A summary of the benefit profiles including key information regarding their attributes
Written by . . .	Program manager	
Approved by . . .		The register is a document used by the program team as a reference
Contents	Reference number for the benefit	
	Benefit name	
	Forecast value of the benefit	
	Measured baseline metrics	
	Timeframe for benefit realization	
	Dependencies • Project • Transition activities • Other benefits	
	Benefit owner	Often the BCM
	Date identified	
	Date realized	
	Measurement approach	
	Notes	

Benefits Map (or other diagrammatic representation)

Purpose	Clarify and record the connections and dependencies between the projects and the benefits	Present a single diagram to stakeholders to describe the intent of the initiative
Written by . . .	Program manager	
Approved by . . .	Sponsor	Often presented with the business case document
Contents	Benefits and dis-benefits	
	Outputs	
	Outcomes	
	Strategic objectives	
	Dependencies within the network	
	Notes regarding: • External dependencies • Major risks • Comments regarding the sources of the information compiled • Business changes required for the sustainability of the benefits	

Business Case

Purpose	Present the information to demonstrate the viability of the initiative—used as the mechanism to make investment decisions	A "live" document which will be updated as the initiative progresses
Written by . . .	Program manager	
Approved by . . .	Sponsor	
Contents	Costs of delivery	Project and program costs
	Cost of transition activities	
	Costs associated with establishing baselines and measuring benefits	
	Timeframe for delivery of the initiative	The overall duration of the initiative and the period of benefits realization, including significant milestones and constraints recorded
	Benefits, including their forecast magnitude and the timing of their realization	
	Risks	A list of the major risks and a summary of their impacts
	Investment appraisal	Comparison of the costs and benefits

Benefits Realization Plan

Purpose	Record the timeframe and schedule of points for measuring the benefits	
Written by . . .	Program manager	With input from BCMs
Approved by . . .	Sponsor	
Contents	Details of individual benefits	Benefit profiles
	Responsibilities for the identification and definition of the benefits	
	Review points for the review of plans and business case documents	
	Resources required to measure the benefits	Internal and external
	Reporting requirements	
	Relationships between the benefits and their connections to the projects	A benefits map may be included to explain these links

Program Plan

Purpose	Records the key information regarding the initiative, with details covering how the initiative will be controlled and managed	The key reference document for the whole initiative
Written by . . .	Program manager	With input from project managers
Approved by . . .	Sponsor	
Contents	Overall schedule for the initiative	
	Details of the projects selected	Includes an explanation of the projects selected and their groupings and associated activities
	Dependencies between projects and activities	
	External dependencies and decisions from outside the initiative	
	Major milestones	To measure progress and make investment decisions
	Monitoring and reporting requirements	
	Costs and resource requirements	
	Major risks and issues	Direction regarding their management (which may take the form of a risk management plan and issue/change management plan)
	Assumptions and lessons which were applied in developing the plan	

Transition Plan

Purpose	Detail all of the activities, resources, and costs associated with the work required to integrate the outputs into the operational environment	
Written by...	BCMs in collaboration with program manager	
Approved by...	Sponsor	
Contents	Triggers for the plan to be enacted	
	Overall schedule for the transition	Detailed timeline for the transition activities
	Dependencies	Between the: • Projects and the transition activities • Operational activities and the transition
	Constraints	Applied to the transition period and activities
	Major risks and issues	Those which specifically address the transition
	Costs and resource requirements	Required for the implementation of the transition
	Assumptions and lessons	To provide an understanding of the context of the plan

Sustainment Plan

Purpose	To create visibility of and a need for action in the wider environment, to ensure that the benefits and the new "business as usual" are sustainable	
Written by ...	Program manager	In consultation with the sponsor and BCMs
Approved by ...	Sponsor	And delegated to the appropriate corporate people or groups
Contents	Triggers for the plan to be enacted	
	Justification for the actions and the plan	Reasoning behind the need for the sustainment activities
	Overall schedule	The timeframe required to embed the changes for long-term sustainability
	Explanation of the impact of the "do nothing" scenario	
	Costs and resource requirements	
	Dependencies	Links among the sustainment activities, the benefits, and corporate objectives.
	Constraints	In terms of timeframe and resources
	Assumptions and lessons	

Review Report

Purpose	Record the progress and success of the initiative at a particular point in time	Reviews can be applied to several elements of the initiative, which may lead to separate templates being produced or some sections being not applicable for each review
Written by . . .	Program manager or reviewers	
Approved by . . .	Sponsor	
Contents	Date of review	
	Participants in the review	
	Timing and purpose of review	With respect to the benefits life cycle—will provide context
	Measured value of benefits	Either as a cumulative figure or as a list of individual benefits
	Planned value of benefits	
	Comparison of actual versus planned benefits	Including a discussion of the differences between the planned and actual results
	Anticipated results	Explanation of deviations from the benefits realization plan due to performance to date
	Business case review	The status of the business case should be reported, with explanation of current or projected shortfalls or overruns
	Recommendations	Recommendations for future actions based on performance to date
	Acceptance	Signed by the sponsor to include instructions and approvals for remedial actions

Benefits Closure Report

Purpose	Define and record the ground rules for the management of benefits within the initiative	"What" and "how" benefits will be approached in the initiative
Written by . . .	Program manager	Generally with consultation with the sponsoring group
Approved by . . .	Sponsor	
Contents	Date of review	
	Participants in the review	
	Measured value of benefits	
	Attributable future benefits	The value and substantiation of future benefits which may be claimed by this initiative
	Planned value of benefits	The anticipated value of the benefits to the end of the benefits life cycle
	Comparison of actual versus planned benefits	A discussion of the differences between the planned and actual results
	Business case review	The final status of the business case should be reported, with explanations of shortfalls or overruns
	Recommendations	Follow-on actions should be recorded, including: • Closure of the initiative • Additional wok and activities to enhance benefits • Sustainment activities
	Lessons learned	A summary of lessons learned should be presented for consideration and dissemination
	Acceptance	The report should be signed off by the sponsor, indicating acceptance of the report and decisions on any recommended actions (if applicable) made regarding follow-up actions and the recommendations

Appendix II

Summary of Cognitive Biases Impacting Benefits Realization Management

Cognitive Biases

Explanation	Remedies—Actions to Combat the Bias
Strategic Bias	
Deliberate adjustment of estimates to be more appealing to stakeholders	• Reference class forecasting • Wisdom of crowds—Involve stakeholders from diverse backgrounds • Seek actual performance metrics from similar projects
Optimism Bias	
The tendency to be overly optimistic and focus on favorable outcomes; underestimating costs and duration of delivery and exaggerating the benefits	• Reference class forecasting • Wisdom of crowds—Involve stakeholders from diverse backgrounds • Independent reviews of estimates • Seek confirmation and justification for estimates
Planning Fallacy	
The duration to complete a future task displays an optimism bias and underestimates the time required—leading to cost overruns and benefit shortfalls	• Reference class forecasting • Wisdom of crowds—The fallacy refers to one's own activities; external observers tend to be more realistic/pessimistic • Independent reviews of estimates • Risk-based review of estimates

(Continued on following page)

Cognitive Biases *(Continued)*

Explanation	Remedies—Actions to Combat the Bias
Bandwagon Effect	
Adopting a position which aligns with consensus and increasingly popular stances	• Reference class forecasting • Wisdom of crowds—The fallacy refers to one's own activities; external observers tend to be more realistic/pessimistic • Independent reviews of estimates
Groupthink	
A tendency for individuals within a group to moderate their views toward a "norm"	• External facilitation • Wisdom of crowds—Involve stakeholders from diverse backgrounds • Seek actual performance metrics from similar projects • Independent reviews of estimates
Abilene Paradox	
When a group agrees to actions which none of the group want to take	• External facilitation
Confirmation Bias	
When information is actively sought to confirm assumptions and preconceived ideas	• Reference class forecasting • Wisdom of crowds—Involve stakeholders from diverse backgrounds • Independent reviews of estimates • Seek confirmation and justification for estimates • Actively seek information which is contrary to estimates
Availability Bias	
Undue importance is placed on information which is readily and easily available; information which is more difficult to obtain is deemed less valuable and valid	• Reference class forecasting • Wisdom of crowds—Involve stakeholders from diverse backgrounds • Independent reviews of estimates • Actively seek information which is contrary to estimates

(Continued on following page)

Cognitive Biases (Continued)

Explanation	Remedies—Actions to Combat the Bias
Loss Aversion	
Losses are rated as having greater impact than an equivalent gain—that is, dis-benefits are deemed to be worse than benefits	Becomes more significant as the initiative progresses, because the dis-benefits often become more apparent as the work is undertaken • Wisdom of crowds—Involve stakeholders from diverse backgrounds • Independent reviews of estimates • Invert the benefits—that is, money will be lost if the initiative is *not* undertaken
Sunk-Cost Effect	
The belief that too much money has already been committed to the initiative so that there is no option to continue	• Plan the initiative in stages • Create a culture where value and benefits are seen as the result of investments • Justify continuation at the end of stages—rather than try to justify stopping • Independent review of the business case
Endowment Effect	
Greater value is placed on assets which are owned by the performing organization, and less value is assigned to assets which are not owned—Manifests as an overvaluation of the status quo and an undervaluation of the capability and future benefits	• Reference class forecasting • Wisdom of crowds—Involve stakeholders from diverse backgrounds • Independent reviews of estimates • Actively seek information which is contrary to estimates
Affect Heuristic	
When there is a known preferred solution, estimates, risks, and plans are presented in such a way as to favor that preferred solution	• External facilitation • Reference class forecasting • Wisdom of crowds—Involve stakeholders from diverse backgrounds • Independent reviews of estimates • Actively seek information which is contrary to estimates
Regression to the Mean	
Assuming that the initial results, particularly if they are greater than forecasts, will continue to be realized at the accelerated level	• Reference class forecasting • Wisdom of crowds—Involve stakeholders from diverse backgrounds • Independent reviews of estimates

(Continued on following page)

Cognitive Biases (Continued)

Explanation	Remedies—Actions to Combat the Bias
Framing	
The presentation of information and options is made in such a manner as to prefer one of the options	• External facilitation • Reference class forecasting • Wisdom of crowds—Involve stakeholders from diverse backgrounds • Independent reviews of estimates
Anchoring	
Disproportionate credibility is given to the first estimate produced	• External facilitation • Reference class forecasting • Wisdom of crowds—Involve stakeholders from diverse backgrounds • Independent reviews of estimates
Rule Beating and Unintended Consequences	
The changed operational environment may give rise to behaviors which are contrary to the intention of the investment	• Reference class forecasting • Wisdom of crowds—Involve stakeholders from diverse backgrounds • Independent reviews of estimates

Abbreviations and Acronyms

APM	Association for Project Management, www.apm.org.uk
BA	Business analysis
BAU	Business as usual
BCM	Business change manager
BCR	Benefit–cost ratio
BDN	Benefits dependency network
BRM	Benefits realization management
CCB	Change control board
CEDEFOP	European Centre for the Development of Vocational Training
CMI	Change Management Institute, www.change-management-institute.com
CPI	Consumer Price Index
CSR	Corporate social responsibilities
DDT	dichlorodiphenyltrichloroethane
EGAP	Everything Will Go According to the Plan
FEL	Front-end loading
FV	Future value
GI Hub	Global Infrastructure Hub
GIA	Governance Institute of Australia
G20	The Group of Twenty—a forum for the governments and central bank governors from the largest global economies
KPI	Key performance indicator
LNG	Liquefied natural gas
NPV	Net present value

OGC	Office of Government Commerce
PESTLE	Political, Economic, Social, Technological, Legal, and Environmental
PgMO	Program management office
PMBOK® Guide	A Guide to the Project Management Body of Knowledge
PMI	Project Management Institute, www.pmi.org
PMO	Project management office
PRINCE2®	PRojects IN Controlled Environments version2—a project management methodology
PV	Present value
PwC	PricewaterhouseCoopers
ROI	Return on investment
TBL	Triple Bottom Line
WHO	World Health Organization

References

APM, 2012, *APM Body of Knowledge, 6th Edition*, Association for Project Management.
APM Benefits Management Special Interest Group, 2011, "Delivering benefits from investments in change: Winning hearts and minds," APM Benefits Management Special Interest Group, Association for Project Management, https://www.apm.org.uk/media/1187/winning-hearts-and-minds.pdf, retrieved March 2018.
AXELOS, 2011, *Managing Successful Programmes, 4th Edition*, TSO (The Stationary Office).
AXELOS, 2017, *Managing Successful Projects with PRINCE2®*, TSO (The Stationary Office).
Baron, J., 1994, *Thinking and Deciding*, 2nd ed., Cambridge University Press, pp. 224–228.
Bennis, W., & Nanus, B., 1985, *Leaders: The Strategies for Taking Charge*, Harper & Row.
Boushey, H., & Glynn, S. J., 2012, "There are significant business costs to replacing employees," Center for American Progress, https://www.americanprogress.org/issues/economy/reports/2012/11/16/44464/there-are-significant-business-costs-to-replacing-employees, retrieved March 2018.
Bradley, G., 2006, *Benefits Realization Management*, Gower.
Bradley, G., 2010, *Fundamentals of Benefits Realization*, TSO (The Stationary Office).
Brooks, F. P., Jr., 1975, *The Mythical Man-Month: Essays on Software Engineering*, Addison-Wesley.
CEDEFOP, 2011, *Research Paper 10: The Benefits of Vocational Education and Training*, Publications Office of the European Union, Luxembourg.
Ciccarelli, S., & White, J., 2014, *Psychology*, 4th ed., Pearson Education, p. 62.
CMI, 2013, *The Change Management Body of Knowledge, 1st Edition*, The Change Management Institute.
Collins English Dictionary, www.collinsdictionary.com, retrieved March 2018.
Cooke-Davies, T., 2002, "The 'real' success factors on projects," *International Journal of Project Management*, 20:185–190.
Covey, S. R., 1989, *The 7 Habits of Highly Effective People*, The Free Press.
Dictionary.com, http://www.dictionary.com/browse/assurance, viewed 4 October 2017.
Dolan, K. J., 2002, "Goal-setting and interactive planning: Improving productivity on construction sites through improved management," University of Dundee.
Drucker, P. F., 2001, *The Essential Drucker*, HarperCollins.
Duff, A., Robertsen, I., Phillips, R., and Cooper, M., 1994, Improving safety by modification of behavior, *Construction Management and Economics*, 12(1):67–78.

Duncan, B., 2015, "Double your project productivity: Why 1 + 1 = 1," LinkedIn, 11 July 2015.

Edkins, A., Geraldi, J., Morris, P., & Smith, A., 2013, Exploring the front end of project management, *Engineering Project Organization Journal*, 3(2):71–85.

Elaurant, S., & McDougall, W., 2014, "Politics, funding and transport—The need for systematic reform," AITPM 2014 National Conference, Adelaide, Australia, August 12–15, 2014.

Elkington, J., 1997, *Cannibals with Forks: The Triple Bottom Line of 21st Century Business*, Capstone.

Flyvbjerg, B., 2003, *Megaprojects and Risk: An Anatomy of Ambition*, Cambridge University Press.

Flyvbjerg, B., 2014, "What you should know about megaprojects and why: An overview," *Project Management Journal*, April/May, p. 12.

Flyvbjerg, B., 2016, "The fallacy of beneficial ignorance: A test of Hirschman's hiding hand," *World Development*, 84:176–189.

Flyvbjerg, B., 2018, "Planning fallacy or hiding hand: Which is the better explanation?" *World Development*, 103:383–386.

Flyvbjerg, B., & Budzier, A., 2015, "Why do projects fail?" *Project Magazine*, Summer, p. 22.

GI Hub, 2017, "Forecasting infrastructure investment needs and gaps," https://outlook.gihub.org/Global Infrastructure Hub, viewed 8 January 2018

GIA, 2017, "What is governance?" Governance Institute of Australia, https//www.governanceinstitute.com.au/resources/what-is-governance, retrieved March 2018.

Harvey, J. B., 1974, "The Abilene paradox: The management of agreement," *Organizational Dynamics*, 3:63–80.

Harvey, J., 1988, "The Abilene paradox: The management of agreement," *Organizational Management*, 17(1):19–20.

Hastie, R., & Dawes, R. M., 2001, *Rational Choice in an Uncertain World*, Sage Publications, p. 288.

Hirschman, A. O., 1967, *Development Projects Observed*, The Brookings Institution.

Jenner, S., 2012, "Benefits realization—Building on (un) safe foundations or planning for success?" *PM World Journal*, 1(1):1–6.

Jenner, S., 2014, *Managing Benefits, 2nd Edition*, APMG International.

Kahneman, D., 2011, *Thinking, Fast and Slow*, Allen Lane.

Kahneman, D., & Tversky, A., 1973, "On the psychology of prediction," *Psychological Review*, 80:237–251.

Kaplan, J., 2005, *Strategic IT Portfolio Management*, PRTM.

Kotter, J. P., 1996, *Leading Change*, Harvard Business School Press.

Kruger, J., & Dunning, D., 1999, "Unskilled and unaware of it: How difficulties in recognizing one's own incompetence lead to inflated self-assessments," *Journal of Personality and Social Psychology*, 77(6):1121–1134.

Landsberger, H. A., 1958, *Hawthorne Revisited*, The New York State School of Industrial and Labor Relations.

Letavec, C. J., 2006, *The Program Management Office—Establishing, Managing and Growing the Value of a PMO*, J. Ross Publishing.

Levitt, S. D., & Dubner, S. J., 2005, *Freakonomics: A Rogue Economist Explores the Hidden Side of Everything*, William Morrow.

Levitt, S. D., & Dubner, S. J., 2016, *When to Rob a Bank . . . and 131 More Warped Suggestions and Well-Intentioned Rants*, William Morrow.

Lovallo, D., & Kahenman, D., 2003, "Delusions of Success: How Optimism Undermines Executives' Decisions," *Harvard Business Review*, July 2003.

Lovins, A., 1977, *Soft Energy Paths: Toward a Durable Peace*, Ballinger.

Mayo, E., 1949, *The Social Problems of an Industrial Civilization*, Routledge & Kegan Paul.
McAllister, I., & Studlar, D. T., 1991, "Bandwagon, underdog, or projection? Opinion polls and electoral choice in Britain, 1979–1987," *The Journal of Politics*, 53(3):720–740.
Meadows, D., 2008, *Thinking in Systems: A Primer*, Chelsea Green Publishing.
Mehrabian, A., 1998, "Effects of poll reports on voter preferences," *Journal of Applied Social Psychology*, 28(23):2119–2130.
Merrow, E. W., 2011, *Industrial Megaprojects: Concepts, Strategies, and Practices for Success*, Wiley.
OGC, 2003, *Gateway News*, December.
OGC, 2011, *Management of Portfolios*, TSO (The Stationary Office).
Orwell, G., 1949, *Nineteen Eighty-Four: A Novel*, Harcourt, Brace & Co.
PMI, 2016a, *Strengthening Benefits Awareness in the C-Suite*, PMI Thought Leadership Series, Project Management Institute.
PMI, 2016b, "The strategic impact of projects: Identify benefits to drive business results," *Pulse of the Profession®* Report, Project Management Institute.
PMI, 2017a, *A Guide to the Project Management Body of Knowledge (PMBOK® Guide), 6th Edition*, Project Management Institute.
PMI, 2017b, *The Standard for Program Management, 4th Edition*, Project Management Institute.
PMI, 2017c, *The Standard for Portfolio Management, 4th Edition*, Project Management Institute.
PMI, 2018, "Success in disruptive times," *Pulse of the Profession®* Report, Project Management Institute.
PwC, 2014, "Capital projects and infrastructure spending; Outlook to 2025," Pricewaterhouse Coopers, www.pwc.com/cpi-outlook2025.
Reich, B. H., 2004, "Knowledge 'traps' in IT projects," Project Management World Today, retrieved from http://pmforum.org/library/papers/2004/1112papers
Saket, M. M., 1986, "Cost significance applied to estimating and control of construction projects," PhD thesis, University of Dundee.
Schueler, J., Stanwick, J., & Loveder, P., 2017, *A Framework to Better Measure the Return on Investment from TVET*, National Centre for Vocational Education Research, https://www.ncver.edu.au/research-and-statistics/publications/all-publications/a-framework-to-better-measure-the-return-on-investment-from-tvet, retrieved January 2018.
Siebert, H., 2002, *Der Kobra-Effekt. Wie man Irrwege der Wirtschaftspolitik vermeidet*, Deutsche Verlags-Anstalt.
Simonson, I., 1989, "Choice based on reasons: The case of attraction and compromise effects," *Journal of Consumer Research*, 16(2): 158–174.
Sims, R. R., 1994, *Ethics and Organizational Decision Making: A Call for Renewal*, Greenwood Publishing.
Surowiecki, J., 2005, *The Wisdom of Crowds*, Anchor Books.
Vann, M. G., 2003, "Of rats, rice, and race: The great Hanoi rat massacre, an episode in French colonial history," *French Colonial History*, 4:191–203.
Ward, J., & Daniel, E., 2006, *Benefits Management—Delivering Value from IS and IT Investments*, Wiley.

Index

A

Abilene paradox, 116, 117
addressing failure, 211
affect heuristic, 121, 122
analysis, 7, 9, 10, 37, 52, 59, 70, 83, 98, 110, 113, 118, 124, 135, 163, 233
anchoring, 123, 124
APM. *See* Association for Project Management
assess the benefits, 66, 109, 121, 146, 147, 211
Association for Project Management (APM), 5, 9
assurance, 26, 48, 56–58, 60, 136, 142–144, 146, 156, 175, 203, 220, 240, 241
assurance roles, 56, 57
attribution, 234, 241
availability bias, 114, 118, 124

B

bandwagon effect, 116
baseline, 7, 33, 34, 39, 77, 81, 102, 103, 106, 110, 111, 140, 157, 158, 188, 189, 196–198, 203, 205, 206, 214, 215, 217, 218, 232, 238
base rate fallacy, 112
BAU. *See* business as usual
BCM. *See* business change manager
BCR. *See* benefit–cost ratio
begin with the end in mind, 86, 87
Ben Franklin effect, 112
benefit, 3–26, 28, 29, 31–42, 45, 47–50, 52–56, 59, 60, 65–67, 69–82, 85–107, 109–115, 118–124, 126, 128–132, 135–155, 158–160, 162–169, 172–175, 177–181, 183–189, 192, 194–201, 203–224, 227–241
benefit cost analysis, 37
benefit–cost ratio (BCR), 37, 38, 114, 129, 136, 137, 147, 148
benefit map, 38, 88, 89
benefit profile, 32, 39, 40, 81, 102–104, 106, 141, 142, 146, 177, 180, 203, 218, 228, 239, 241
benefit realization plan, 40, 41, 177
benefit review, 155, 209
benefits closure report, 42, 220–222
benefits coordinator, 59. *See also* benefits manager

benefits dependency map, 94, 95, 97
benefits dependency network, 38, 92, 93, 97
benefits life cycle, 10, 28, 42, 47, 60, 65–67, 69, 70, 85, 109, 121, 145, 149, 150, 168, 183, 184, 204, 207, 209, 211, 213, 216, 218, 220–223, 228, 236
benefits logic map, 95–97
benefits management life cycle, 35
benefits management strategy, 39, 42, 80, 82, 189, 210, 219, 220, 234–236
benefits manager, 59
benefits realization management (BRM), 5, 9, 10, 16, 21, 25, 26, 49, 59, 66, 79, 98, 103, 121, 150, 152, 179, 223, 227–233, 235–238, 240–242
benefits realization plan, 39, 49, 66, 81, 140, 153, 155, 166, 179, 204, 210, 212, 217, 222
benefits realization strategy, 23, 141, 223
benefits register, 40, 103, 104, 142, 218
BRM. *See* benefits realization management
business analysis, 52, 59
business as usual (BAU), 28, 31, 35, 138, 140, 153, 158, 191, 193, 194, 196, 198
business case, 6, 16, 18–20, 26, 31, 32, 40, 42, 47, 48, 50, 56, 57, 66, 72–74, 78, 81, 86, 87, 101, 110, 111, 114, 119, 129, 135, 137, 138, 140–144, 146, 147, 161, 162, 166, 168, 169, 174, 175, 177, 180, 185–188, 195, 203, 204, 209, 210, 216–218, 221, 222, 228, 232–236, 238

business case review, 209, 222
business change, 9, 36, 38, 41, 45, 48–50, 53–55, 57, 90, 92, 94, 95, 100–102, 104, 105, 145, 146, 152, 178, 184, 211, 217
business change manager (BCM), 36, 41, 45, 48–50, 53–57, 90, 104, 145, 146, 152, 153, 158–165, 175, 178, 184–187, 191–195, 197–199, 203, 204, 211, 217
business interest, 47

C

capability, 9–12, 28, 30, 31, 47, 50, 74, 79, 89, 90, 115, 160, 186, 192, 199, 239, 241
champions, 240, 241
change, 4–9, 13, 14, 17, 20, 21, 23, 25, 27, 28, 30–34, 36–41, 43, 45, 47–55, 57–61, 66, 67, 70–73, 75, 77–79, 83, 86–90, 92, 94, 95, 97–106, 110, 111, 115, 116, 119, 120, 122, 123, 126, 129, 132, 136–138, 140–143, 145–147, 150–153, 155, 157–169, 174, 175, 177–180, 184–204, 208–211, 213, 215–219, 222, 228–231, 236–239, 241
change control board, 59
Change Management Institute (CMI), 5
change team, 55, 59, 61, 180, 199, 203
change the conversation, 23, 228, 241
characteristics of benefits, 6
closure, 26, 42, 210, 219–222
CMI. *See* Change Management Institute
cobra effect, 125, 126
cognitive bias, 111, 112

Index 267

combatting the biases, 126
commission, 8, 50, 77, 132, 147
communication, 10, 19, 21, 38, 41, 45, 47, 53, 55, 80, 120, 122, 151, 158, 162, 164, 165, 168, 178, 179, 185–188, 192, 199, 203, 205, 234–236, 239
compliance, 9, 57–59, 70, 75, 77, 78, 83
compliance initiatives, 75, 77, 83
compliance projects, 78
concurrent projects, 74, 169, 172, 173
confirmation bias, 114, 117, 118, 124
consolidation, 60, 77, 174, 200
consulting organization, 52
context, 4, 16, 27, 34, 39, 40, 42, 47, 49, 66, 67, 69, 70, 73, 75, 77, 80–82, 86, 87, 98–100, 111, 116, 121, 128, 140, 141, 156, 157, 163, 184, 189, 192, 221, 230, 232, 236
contractual and legal, 94
coordinate and realize the benefits, 66, 183, 203, 209, 211, 238
Corporate Social Responsibilities (CSR), 16
courtesy bias, 112
CSR. *See* Corporate Social Responsibilities
culture, 11, 30, 77, 94, 174, 229, 230

D

decision making, 35, 48, 58, 59, 70, 79, 110, 111, 118, 120, 122, 133, 141, 145, 147, 151, 197, 208, 219, 227, 230, 235, 239
decommission, 161, 198, 203, 205
deliverable, 5, 10, 28, 45, 79, 97, 155, 156, 159, 191–193, 198, 204
delivery, 4, 6, 9, 10, 13, 20, 21, 23, 25, 26, 28, 30, 34, 35, 37–40, 45, 47, 49–51, 56, 58, 66, 76, 77, 79, 81, 86, 87, 90, 95, 101, 102, 105, 120, 122, 124, 128, 136, 137, 144, 150–153, 155, 158, 162, 164, 165, 168, 169, 175–179, 184, 190, 208, 214, 215, 218, 220, 227, 229, 235, 241
Delphi technique, 128
dependency, 11, 38, 41, 59, 66, 87, 90, 92–95, 97, 101, 103–105, 138, 153, 168, 169, 177–179
dependency network, 38, 92, 93, 97
description, 7, 33, 47, 61, 102, 157
diagrammatic approaches/techniques, 87, 88, 92, 97
dis-benefits, 28, 32, 94, 104–106, 110, 119, 122, 138, 187, 200, 201, 213, 233, 241
discount rate, 129–132, 148, 166
double dipping, 235, 236, 241
driver, 32, 70–72, 75, 80, 83, 95, 100, 160, 195, 219, 236
Dunning–Kruger effect, 112

E

early wins, 166–169, 171, 174, 181, 230, 238
economics, 16, 70–73, 83, 127, 132, 135, 148
effectiveness, 13, 16–18, 20, 205, 230
efficiency, 7, 13, 16–18, 20, 32, 217
EGAP. *See* Everything Will Go According to the Plan
emergent benefits, 28, 31, 32, 67, 77, 155, 163, 196, 211, 217, 218
emergent programs, 86, 90
enabler, 38, 78, 92, 95

endowment effect, 121
environment, 4–6, 9–11, 13, 16, 21, 23, 26–28, 30, 31, 33–35, 41, 45, 47, 51–55, 60, 66, 67, 70, 72–79, 81–83, 86, 89, 90, 92, 95, 98, 101, 102, 106, 112, 116, 119, 122, 124, 142, 143, 148, 150–153, 155, 158, 159, 161, 162, 167, 168, 178, 181, 184–191, 193, 194, 198–203, 205, 212, 217, 219, 227–230, 236, 240, 241
environmental benefits, 16, 74
establish the context, 66, 69, 70, 82
estimate, 4, 14, 32, 111–115, 119, 121, 123, 124, 126–128, 132, 143, 144, 148, 151, 152, 156, 169, 212, 215–217
Everything Will Go According to the Plan (EGAP), 114, 115
evolving initiative, 75–77, 83
exception, 47, 49, 51, 189, 210

F

failure, 5, 9, 17, 72, 87, 97, 114, 115, 145, 158, 161, 164, 197, 211, 212
FEL. *See* front-end loading
financial year, 36, 135
fiscal year, 7, 36, 125, 130, 166, 212
forecast, 3, 7, 8, 28, 31, 42, 47, 55, 56, 79, 87, 90, 111, 114, 115, 117, 120–122, 127, 131, 140, 142, 143, 148, 155, 157, 160, 162, 163, 166, 174, 185, 187, 188, 198, 208, 209, 211–216, 218, 220, 224
framing, 123
front-end loading (FEL), 36, 50, 175, 181

G

gateway reviews, 47, 56
gateways, 35, 36, 47, 56, 142, 144
GI Hub. *See* Global Infrastructure Hub
Global Infrastructure Hub (GI Hub), 3, 4
governance, 17, 20, 26, 32, 35, 42, 48, 57, 58, 60, 75, 83, 145, 151, 156, 157, 227
governance roles, 57, 58
groupthink, 116–118, 128, 152

H

Hawthorne effect, 188

I

identify the benefits, 38, 66, 85, 88, 104, 105, 110, 211
independent review, 144, 148
induction, 14, 30, 33, 161, 165, 166, 191, 198, 203, 227, 231, 242
initiative, 6, 8, 9, 13, 15–21, 23, 25, 27, 28, 30–45, 47–51, 53–61, 66, 67, 70–83, 86, 88, 90, 92, 95, 97–101, 106, 111–113, 120, 121, 131, 135, 142, 144, 145, 147, 148, 150, 153, 155, 158, 162–164, 166–169, 174–178, 180, 181, 184, 186, 188, 206–209, 211, 217–223, 228–236, 238–241
intangible, 13, 18–20, 103
integration, 174, 184, 198, 227, 230
intermediate benefits, 11, 28, 31, 87, 89, 90, 95, 106
investment, 4, 5, 8, 9, 13–16, 18, 19, 21, 23, 26, 27, 35–42, 47, 48,

50, 52, 57, 58, 60, 66, 70–75, 77, 78, 83, 86, 88, 90, 92, 95, 100, 101, 112–114, 120, 129–133, 135–137, 140–142, 144, 147, 148, 152, 162, 166, 173, 176, 184, 186, 188, 189, 195, 208, 210, 217, 219, 227–232, 234–236, 240, 241
investors, 4, 79, 101, 106, 113, 135, 136, 209, 221

J

journey, 10, 11, 88, 117

K

knowledge organization, 52
Kotter's change model, 119, 237

L

learning lessons, 105, 145
legitimacy, 7, 8, 218, 235, 241
loss aversion, 119, 120
lost benefits, 224

M

measuring benefits, 174, 196, 234
model for change, 236, 241

N

net present value (NPV), 37, 129–133, 136, 147, 148, 166, 169
new operating environment, 28, 158, 162
New Royal Adelaide Hospital, 159, 165
NPV. *See* net present value

O

objective, 3–6, 8–11, 19, 21, 23, 26–28, 31, 38, 49, 70, 74–80, 82, 83, 86–90, 92, 95, 97, 98, 100, 104–107, 110, 120, 132, 136, 137, 143, 144, 148, 157, 163, 176, 179, 192, 206, 212, 219, 221, 228, 229, 233, 235, 236
observable, 7, 33, 140, 199
Office of Government Commerce (OGC), 8, 27
OGC. *See* Office of Government Commerce
operational change, 97, 98, 110, 153, 201
operational environment, 4–6, 11, 13, 23, 26–28, 30, 31, 33–35, 41, 45, 51, 53–55, 60, 66, 79, 86, 89, 90, 92, 101, 102, 151, 152, 155, 158, 161, 167, 168, 178, 184–186, 189, 190, 193, 198–200, 205, 236
optimism bias, 114, 120
outcome, 6, 11, 12, 17, 18, 26, 28, 31–33, 38, 45, 70, 72, 75–78, 87, 89, 90, 92, 103–105, 114–116, 118, 121, 123, 126–128, 138, 145, 152, 153, 155, 158–162, 166, 174, 184, 187, 188, 191–194, 198, 202, 203, 205, 209, 218, 220
outputs, 9–13, 17, 20, 21, 28, 30, 31, 33, 41, 51, 79, 89, 90, 92, 99, 105, 152, 178, 184, 199, 205, 241

P

payback period, 37, 129, 133, 135, 148, 173

PESTLE. *See* Political, Economic, Social, Technological, Legal, and Environmental
PgMO. *See* Program Management Office
plan for benefits realization, 66, 149
planning fallacy, 114, 115, 120
PMI. *See* Project Management Institute
PMO. *See* Project Management Office
Political, Economic, Social, Technological, Legal, and Environmental (PESTLE), 70, 71, 83
portfolio management office, 175, 236
post-transition, 162, 184, 189, 190, 194–198, 200, 202, 205, 209, 219
present value (PV), 37, 129, 130, 147, 166, 169
pre-transition, 184, 185, 192, 198, 200, 205
PRINCE2®, 26, 47, 145
procedure, 12, 30, 94, 227
procurement, 59, 156
product, 4, 13, 16, 17, 20, 21, 25, 26, 28, 30, 31, 51, 54, 66, 79, 90, 99, 101, 102, 124, 158, 184–186, 190, 191, 203, 216
productivity, 13, 14, 17, 18, 20, 33, 86, 99, 107, 141, 176, 187, 188, 196, 200, 213
Program Management Office (PgMO), 23, 49, 51, 52, 58, 175, 210
program manager, 45, 48–51, 53, 54, 57, 60, 70, 72, 80–83, 90, 98, 101, 104, 145, 146, 152, 153, 176, 178, 180, 195, 198, 203, 204, 208, 210, 218, 220, 222, 223, 232, 236, 238, 239
program office, 49, 52, 53, 58, 59
program plan, 40, 41, 50, 176, 180, 204, 209, 217
program review, 209
progressive elaboration, 27, 28, 32, 150
project, 3–15, 17, 21, 23, 25–28, 30, 32–35, 37, 38, 40, 41, 43, 45, 47, 49–55, 57, 59, 60, 66, 70, 71, 73–81, 83, 86–90, 92, 94, 95, 97, 100–106, 110–116, 118, 120–122, 124, 126–128, 132, 133, 135–138, 140–145, 147, 148, 150–153, 155–161, 163–169, 171–179, 181, 184–187, 189–195, 198–201, 203–205, 208–210, 212–215, 217, 218, 220, 228–230, 233–235, 239, 241
project commissioning, 155
Project Management Institute (PMI), 3, 6, 25–27, 145, 180
Project Management Office (PMO), 23, 51, 52, 144
project office, 50, 52, 53
project plan, 153, 179
project review, 209
proxy measures, 13, 14, 19
PV. *See* present value (PV)

Q

quantification, 9, 52, 110–113

R

ramping, 214, 215

readiness, 31, 41, 56, 159, 163, 205
reference class forecasting, 126, 127
reference group, 152, 200, 201
regression to the mean, 122
reinforcement, 151, 160, 161, 187, 188, 191, 192, 196, 206, 231
reinforcement of change, 160
responsive actions, 163
return on investment (ROI), 9, 23, 36, 73
review, 10, 30, 35, 36, 41, 42, 47, 56–59, 66, 81, 111, 140, 142–144, 146, 148, 152, 155, 156, 159, 162, 165, 167, 174, 177, 180, 194, 197, 201, 204, 206–223, 233, 235, 238, 240, 241
review report, 41, 220–222
review the initiative, 66, 207, 209, 220, 238
right benefits, 98, 99
risk, 30, 36, 49–53, 56–60, 71, 81, 83, 94, 99, 101, 103, 105, 110, 112–115, 121, 122, 128–130, 132, 133, 135, 137, 142, 148, 150, 151, 157, 161, 169, 174, 176–178, 190, 197, 208, 216, 218
risk-based approach, 57, 60
risk manager, 58, 60
ROI. *See* return on investment
role, 7, 10, 14, 23, 30, 40, 42, 43, 45, 47–61, 98, 144, 145, 158, 164, 165, 175, 176, 184, 185, 193, 195, 197, 204, 206, 208, 209, 229–231, 236, 240
rule beating, 125

S

schedule change, 186
scheduling, 52, 53, 59, 127, 156, 169, 174, 196
scheduling and monitoring, 59
scope change, 163, 186
sequencing, 157
significant benefits, 48, 232, 233, 241
SMART, 176, 238, 239, 241
social benefits, 16
solution, 79, 95, 101, 128, 193
specialist support roles, 58
sponsor, 6, 18, 31, 32, 35, 36, 45, 47–51, 54, 57–59, 70, 77, 78, 81, 82, 88, 92, 95, 100, 105, 111–114, 126, 131–133, 136, 137, 140, 146, 157–159, 162, 164, 168, 174–176, 178, 180, 185, 186, 204, 210–212, 217, 219–223, 232, 236, 238–240
sponsoring group, 30, 45–48, 50, 58, 80–82, 99, 104, 113, 145–148, 180, 181, 220, 230–232, 235, 236, 239–241
stage boundaries, 155
stakeholder, 5–10, 13, 15, 16, 18–21, 23, 25, 30–33, 35, 36, 38–40, 42, 43, 45, 47–51, 53, 54, 56, 60, 66, 67, 70–72, 74–83, 86–90, 92, 95, 97–103, 105, 110, 111, 113–115, 118–123, 125, 136–138, 140, 141, 144–146, 151, 153, 155, 157, 158, 160, 162–168, 174, 176, 179, 180, 185–189, 191–195, 197, 201, 203–205, 208, 211–213, 215–218, 221, 228–236, 238–242
stakeholder engagement, 9, 23, 53, 79, 166, 167, 239, 240
standards organization, 51

stepping stones, 31, 89, 194
strategic bias, 112, 240
strategic initiative, 75, 83
strategic objective, 3, 6, 26–28, 86, 88, 97, 105
sunk-cost effect, 113, 120
supplier interest, 47
sustainability, 28, 31, 66, 105, 122, 148, 153, 155, 160, 193, 194, 211, 214, 216, 219, 223
sustainment plan, 41, 66, 153, 155, 178–180, 198, 203, 238

T

tangible, 13, 18–20, 101
TBL. *See* Triple Bottom Line
technology-driven initiatives, 79, 83
training, 20–22, 28, 30, 33, 94, 145, 158, 159, 161, 163, 165, 185–187, 190–194, 198, 201–203, 205, 213
Transbay Terminal, 113 transition, 26, 28, 30–35, 37, 41, 45, 49, 53–55, 66, 89, 103, 110, 111, 136, 151–153, 155, 158, 159, 162, 165, 166, 175, 178–180, 184–194, 196–198, 200–205, 211–216, 223, 235
transition activities, 33, 103, 158, 165, 166, 178, 186, 187, 191, 213
transition period, 28, 31–35, 41, 54, 111, 159, 178, 187, 190–193, 214, 215
transition plan, 41, 66, 153, 155, 158, 178–180, 186, 203, 204
Triple Bottom Line (TBL), 15, 16, 72, 73
triple constraints, 4, 9, 185
type of program, 70, 74, 83

U

underestimate, 112–114
unintended behavior, 125
user interest, 47

V

value, 3, 6–10, 13, 14, 16, 17, 28, 36–38, 40–42, 59, 66, 72, 74–76, 78, 79, 87, 99, 103, 110–113, 116–118, 121, 124, 125, 127–133, 135–137, 141, 147, 148, 150, 160, 162, 166, 167, 169, 173, 175, 184, 186, 190, 195, 208, 211–217, 219–223, 228, 230, 232, 233, 235, 240
vision-led programs, 70

W

wisdom of crowds, 128, 151
work breakdown structure, 156